GEOLOGY AND MINERALOGY RESEARCH DEVELOPMENTS

RESEARCH FRONTIERS OF SEDIMENTARY BASIN INTERIORS

A CASE STUDY AND METHODOLOGICAL REVIEW ON AN OBLIQUE CONVERGENT MARGIN

Library of Congress Cataloging-in-Publication Data

Names: Itoh, Yasuto, editor. | Kusumoto, Shigekazu, editor. | Takemura, Keiji, editor.
Title: Research frontiers of sedimentary basin interiors : a case study and methodological review on an oblique convergent margin / editors, Yasuto Itoh, Shigekazu Kusumoto and Keiji Takemura (Graduate School of Science, Osaka Prefecture University, Osaka, Japan and others).
Description: Hauppauge, New York : Nova Science Publishers, Inc., [2016] | Series: Geology and mineralogy research developments | Includes bibliographical references and index.
Identifiers: LCCN 2016024882 (print) | LCCN 2016027481 (ebook) | ISBN 9781634852500 (softcover) | ISBN 9781634855754 ()
Subjects: LCSH: Sedimentary basins. | Convergent margins. | Plate tectonics. | Subduction zones. | Geology, Structural.
Classification: LCC QE615 .R45 2016 (print) | LCC QE615 (ebook) | DDC 551.44--dc23
LC record available at https://lccn.loc.gov/2016024882

Published by Nova Science Publishers, Inc. † New York

CONTENTS

Preface **vii**

Chapter 1 Basin Formation on Convergent Margins **1**
Yasuto Ito and Shigekazu Kusumoto

Chapter 2 Estimations of Subsurface Structures by Gravity Anomaly and Gravity Gradient Tensor **9**
Shigekazu Kusumoto

Chapter 3 Subsurface Structure of Osaka Sedimentary Basin and Its Tectonic Evolution **27**
Yasuto Itoh

Chapter 4 Tectonic Implication of Basin Formation on the Basis of Paleogeography Using Stratigraphy and Shallow Subsurface Structure from a Drilling Data Base **71**
Keiji Takemura

About the Authors **89**

Index **91**

PREFACE

This book provides readers with an overview of the latest research on sedimentary basins. Although these basins serve as an excellent record of long-term environmental changes, their deep interiors are difficult to investigate because of their harsh pressure and thermal conditions. Controlled by the transition of stress regimes linked to changes in subduction obliqueness, convergent plate margins are the sites of a variety of basin formation processes, the accommodations of which govern the mass balance around the surface of the globe. As a typical example of such a case, the Osaka sedimentary basin, an intra-arc basin in southwest Japan that rapidly developed under neotectonic events on the converging front of the Philippine Sea Plate, is given particularly close consideration in this book. The Osaka sedimentary basin is an active depression that developed at the termination of an arc-bisecting fault and is characterized by complex compartmentalization as a result of slips on numerous secondary faults.

Following an explanation of the regional tectonic context of the case study area in Chapter 1, state-of-the-art methodologies for analyzing gravity anomalies and gravity gradient tensors are presented in Chapter 2, and the subsurface structural features of the tectonic basin are delineated. The high-resolution seismic interpretation and stratigraphic study of deep boreholes discussed in Chapter 3 demonstrate the validity of the basin-wide view suggested by gravitational research, and a timeline of the Quaternary geologic events and causative stress changes is proposed with the support of new radiometric ages obtained from basin-filling sediments. Finally, the tectono-sedimentological synthesis of the basin is presented in Chapter 4 in the form of a paleogeographic reconstruction based on authentic field geology and an intra-basinal geomorphic dissection utilizing a shallow boring database. The former leads to an idea of the regional migration of the depocenter under transient tectonic stress, and the latter provides a closer view of the basin architecture under the influence of eustatic changes in the sea level.

Our vision is that the quantitative methodologies presented in this book and further efforts to integrate scientific knowledge will pave the way for advancements in the comprehension of the diverse development processes of sedimentary basins, which reflect Earth's dynamic evolution.

Chapter 1

BASIN FORMATION ON CONVERGENT MARGINS

Yasuto Itoh [1,*] ***and Shigekazu Kusumoto*** [2]

[1]Graduate School of Science, Osaka Prefecture University, Osaka, Japan

[2]Graduate School of Science and Engineering for Research, University of Toyama, Toyama, Japan

ABSTRACT

Controlled by the obliquity of the subduction of oceanic plates, various types of sedimentary basins develop on convergent margins. The eastern Eurasian margin has been a site of vigorous deformation since the late Mesozoic, reflecting the complicated history of the genesis, migration, and convergence of oceanic plates. Among the countless tectonic zones on the margin, the Osaka sedimentary basin, an intra-arc basin in the eastern part of southwestern Japan, was selected as the focus of the present study. Geophysical research, such as gravity analysis, provides volumetric and morphological insights into neotectonic depressions, and stratigraphic data on deep boreholes provide the time sequence of basin-forming episodes.

[*] Correspondence: yasutokov@yahoo.co.jp.

1. INTRODUCTION

At convergent margins, the continental crust experiences tectonic stress as a result of the subduction of oceanic plates. Part of the resulting strain is released elastically as earthquakes, and the remainder causes permanent crustal depressions, i.e., sedimentary basins. The process of basin formation has long been one of the most significant themes of Earth science because basin-filling sedimentary layers provide an excellent record of long-term environmental changes.

Itoh (2013) provided an overview of basin formation mechanisms at various convergent margins and noted that the obliquity of subduction is an important factor in the mode of crustal deformation. Normal subduction results in the dominance of compressive stress on margins and vigorous arc volcanism and induces convection in the wedge of upper mantle, leading to the eventual opening of the backarc. Oblique subduction activates bisecting faults within arcs and causes the substantial lateral migration of crustal blocks. The prevailing shear stress affects the architecture of basins in forearc regions and controls the formation of barriers to sediment influx.

The history of sedimentary basin development around the Japanese archipelago, which serves as a typical example in the Eurasian margin, since the late Mesozoic was summarized by Itoh (2013) as follows.

1) The prolonged transcurrent faulting from the Late Cretaceous to the Paleogene resulted in the formation of many pull-apart basins on the margin. The propagation of fault terminations caused depocenter migration; wrench deformation was related to the compart-mentalization of the basins.
2) The Neogene was heralded by the emergence of rift zones and the opening of the Japan Sea backarc. It was also characterized by the remarkable rise of a compressive stress regime as a result of multiple arc–arc collision events, which enhanced the formation of foreland and intra-arc basins.
3) The Eurasian margin since the late Neogene has been characterized by the wax and wane of crustal deformation, reflecting a shift in the convergence of the Pacific and Philippine Sea Plates. Compressive episodes resulted in uplift of arc-parallel watershed mountains and the rapid burial of basins by voluminous clastics derived from exhumed hinterlands. Differential crustal motion during shear deformation phases has diversified the basin configuration.

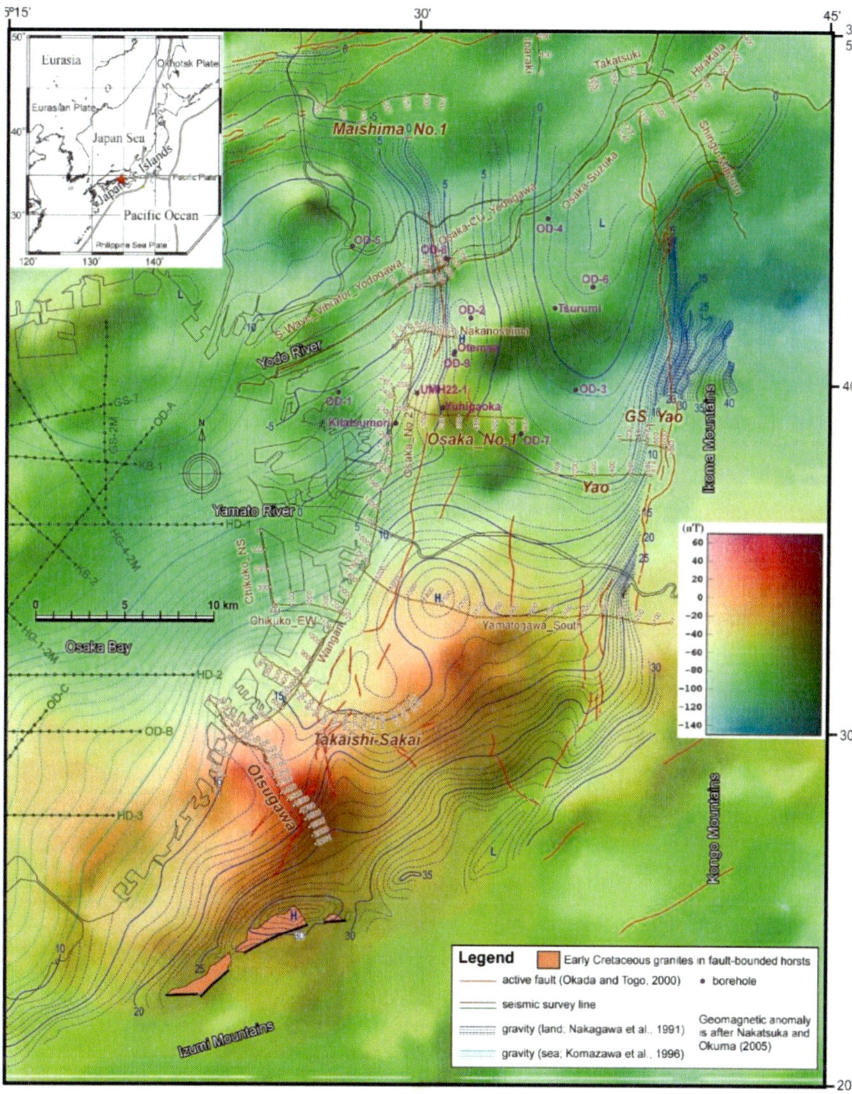

Figure 1. Summary of geophysical and geological data around the Osaka sedimentary basin. The base map shows aeromagnetic anomalies after Nakatsuka and Okuma (2005). The gravity anomaly contours were compiled after Nakagawa et al. (1991) and Komazawa et al. (1996). The red lines show active faults after Okada and Togo (2000). The codes of seismic lines described in this book are highlighted with larger fonts. The inset map depicts the plate tectonic configuration around the Japanese archipelago and the study area (star).

The authors of this book concentrate on the final stage, from the late Neogene to the present, because it has the highest geologic data recovery, enabling the construction and verification of an exquisite tectonic model based on recent geophysical studies, such as the latest gravity analysis and seismic interpretation studies.

2. Tectonic Context of Study Area

2.1. Existing Dataset and Basin Configuration

The focus of this study is the Osaka sedimentary basin, an intra-arc basin in the eastern part of southwestern Japan. Figure 1 shows its geophysical properties (gravity and geomagnetic anomalies) and the existing dataset for basin study (deep boreholes and reflection seismic lines). The Osaka sedimentary basin is a large box-shaped depression surrounded by active faults. Itoh et al. (2013a) estimated the total volume of the basin to be 9.1×10^2 km^3 based on gravity analysis.

The enormous accommodation of the basin is the result of vigorous motions on surrounding faults, which concurrently enhanced the basin compartmentalization. Itoh et al. (2013b) demonstrated that throughout the late Quaternary, the Osaka basin was episodically divided into smaller sags by the differential motion of the crustal blocks on the basis of the compilation of stratigraphic data describing deep boreholes and chronological research.

2.2. Gravity Anomalies of Osaka District

Figure 2 is a residual Bouguer gravity anomaly map of the Osaka district obtained by removing the first-order trend surface (regional linear trend of the Bouguer gravity anomalies) from the Bouguer gravity anomalies. The first-order trend surface was estimated from the Bouguer gravity anomalies using the least squares method. The Bouguer gravity anomaly used here was based on the gravity anomaly database (1 km × 1 km mesh data) by Komazawa (2004), and a Bouguer density of 2670 kg/m^3 was employed.

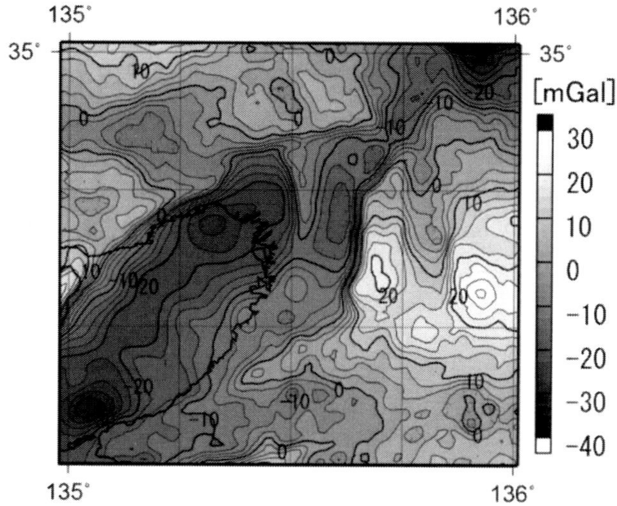

Figure 2. Residual Bouguer gravity anomaly map obtained by subtracting first-order trend surface (regional linear trend of Bouguer gravity anomalies) from Bouguer gravity anomalies. A Bouguer density of 2670 kg/m^3 was employed, and the contour interval is 2.5 mGal. This Bouguer gravity anomaly map was made from the gravity database by Komazawa (2004).

The Osaka district is characterized by a low gravity anomaly with steep north–south-, west–east-, and northeast–southwest-trending horizontal gravity gradient belts. These steep gradient belts correspond to known active fault traces in the area: the north–south belts in the Osaka Plain coincide with the Uemachi Fault and the Ikoma Fault System, and the west–east and northeast–southwest belts around the Osaka Plain correspond to the Arima–Takatsuki Tectonic Line and the Rokko–Awaji Fault System (e.g., Research Group for Active Faults, 1991). In these areas, large density contrasts and/or vertical displacements have been predicted within the surface crust.

Moreover, the Uemachi Highland formed by the Uemachi Fault divides the north–south gravity anomalies in the Osaka Plain into two parts. The eastern part has a north–south trend and is parallel to the Ikoma Fault System. The western part is connected to Osaka Bay, which presents the lowest gravity anomaly in its southern part. The southern part of the Osaka Plain is characterized by two relatively high circular gravity anomalies associated with strong magnetic anomalies, as noted by Itoh et al. (2013a).

Conclusion

This chapter presented an overview of the mechanism of basin formation on convergent plate margins. It has been suggested that the obliquity of the subduction of oceanic plates is a crucial factor in the mode of crustal deformation. The Eurasian margin since the late Neogene has been characterized by the wax and wane of crustal deformation, during which the target Osaka sedimentary basin has been influenced by the shift in the convergence of the Philippine Sea Plate. Vigorous motions on faults surrounding the basin have simultaneously provoked the extensive subsidence and compartmentalization of the tectonic basin. The outline and deep structures of the Osaka basin are nicely delineated by gravity anomaly data.

Acknowledgments

This work was supported partially by JSPS (Japan Society for the Promotion of Science) KAKENHI Grant Number 15K14274.

References

Itoh, Y. (Ed.) (2013). Mechanism of Sedimentary Basin Formation - Multidisciplinary Approach on Active Plate Margins. Rijeka: InTech. http://dx.doi.org/10.5772/50016.

Itoh, Y., Kusumoto, S., & Takemura, K. (2013a). Characteristic basin formation at terminations of a large transcurrent fault - basin configuration based on gravity and geomagnetic data. In Y. Itoh (Ed.), Mechanism of Sedimentary Basin Formation - Multidisciplinary Approach on Active Plate Margins. Rijeka (Croatia): InTech. http://dx.doi.org/10.5772/56702.

Itoh, Y., Takemura, K., & Kusumoto, S. (2013b). Neotectonic intra-arc basins within southwest Japan - conspicuous basin-forming process related to differential motion of crustal blocks. In Y. Itoh (Ed.), Mechanism of Sedimentary Basin Formation - Multidisciplinary Approach on Active Plate Margins. Rijeka (Croatia): InTech. http://dx.doi.org/10.5772/56588.

Komazawa, M. (2004). Gravity grid database of Japan, Gravity CD-ROM of Japan, ver. 2, Digital Geoscience Map P-2 [CD-ROM]. Tsukuba: Geological Survey of Japan.

Komazawa, M., Ohta, Y., Shibuya, S., Kumai, M., & Murakami, M. (1996). Gravity survey on the sea bottom of Osaka Bay and its subsurface structure. *Butsuri-Tansa* (Geophysical Exploration), 49, 459-473.

Nakagawa, K., Ryoki, K., Muto, N., Nishimura, S., & Ito, K. (1991). Gravity anomaly map and inferred basement structure in Osaka Plain, central Kinki, south-west Japan. *Journal of Geosciences,* Osaka City University, 34, 103-117.

Nakatsuka, T., & Okuma, S. (2005). Aeromagnetic anomalies database of Japan, Digital Geoscience Map P-6 [CD-ROM]. Tsukuba: Geological Survey of Japan.

Okada, A., & Togo, M. (2000). Active Faults in Kinki. Tokyo: University of Tokyo Press, 395pp.

Research Group for Active Faults (1991). The Active Faults in Japan: Sheet Maps and Inventories Rev. ed. Tokyo: University of Tokyo Press, 437pp.

Chapter 2

ESTIMATIONS OF SUBSURFACE STRUCTURES BY GRAVITY ANOMALY AND GRAVITY GRADIENT TENSOR

Shigekazu Kusumoto*

Graduate School of Science and Engineering for Research,
University of Toyama, Toyama, Japan

ABSTRACT

In Chapter 2, a short review of semi-automatic interpretation methods for extracting subsurface structure information was conducted, and the horizontal gravity gradient, TDX, Tensor Euler deconvolution and dip angle estimation methods were applied to the gravity gradient tensor obtained from the Bouguer gravity anomaly in and around the Osaka district. As a result, characteristic structures in this region were extracted successfully and it was found that dip angles of active faults were estimated reasonably and without an inconsistency in geological and tectonic viewpoints.

1. INTRODUCTION

In qualitative and/or quantitative analysis of gravity anomaly field, semi-automatic interpretation methods are often employed as a very useful

---

* Correspondence: kusu@sci.u-toyama.ac.jp.

technique for estimating location or edges of causative bodies. These methods have played an important role in rapid and rough analyses of any potential field data as they allow obtaining high objectivity results in a short calculation time without needing geological and geophysical backgrounds.

Aside from including edge detection or edge emphasis methods such as horizontal differential coefficient (e.g., ten Brink et al., 1993; Kusumoto et al., 1996), these techniques also comprise location estimations of causative bodies such as Euler deconvolution (e.g., Thompson, 1982; Reid et al., 1990; Zhang et al., 2000; Rodrigues et al., 2014).

In these semi-automatic interpretation methods, we often employed not only the gravity anomaly but also its vertical and/or horizontal gradients, since the differential coefficients of the gravity anomaly have higher sensitivity than the gravity anomaly itself.

The horizontal and vertical gradients of the gravity anomaly are defined through the gravity gradient tensor Γ. This one is described using the gravitational potential W as (e.g., Torge, 1989; Hofmann-Wellenhof and Moritz, 2005)

$$\Gamma = \begin{bmatrix} \frac{\partial^2 W}{\partial x^2} & \frac{\partial^2 W}{\partial x \partial y} & \frac{\partial^2 W}{\partial x \partial z} \\ \frac{\partial^2 W}{\partial y \partial x} & \frac{\partial^2 W}{\partial y^2} & \frac{\partial^2 W}{\partial y \partial z} \\ \frac{\partial^2 W}{\partial z \partial x} & \frac{\partial^2 W}{\partial z \partial y} & \frac{\partial^2 W}{\partial z^2} \end{bmatrix} \tag{1}$$

Defining g_x, g_y, g_z, as the differentials of W in the x-, y-, and z-directions, we can rewrite equation (1) as

$$\Gamma = \begin{bmatrix} \frac{\partial g_x}{\partial x} & \frac{\partial g_x}{\partial y} & \frac{\partial g_x}{\partial z} \\ \frac{\partial g_y}{\partial x} & \frac{\partial g_y}{\partial y} & \frac{\partial g_y}{\partial z} \\ \frac{\partial g_z}{\partial x} & \frac{\partial g_z}{\partial y} & \frac{\partial g_z}{\partial z} \end{bmatrix} = \begin{bmatrix} g_{xx} & g_{xy} & g_{xz} \\ g_{yx} & g_{yy} & g_{yz} \\ g_{zx} & g_{zy} & g_{zz} \end{bmatrix} \tag{2}$$

Here, g_z is the well-known gravity anomaly and g_{zz} is the vertical gradient of the gravity anomaly. The gravity gradient tensor is symmetric (e.g., Torge,

1989) and the sum of its diagonal components is zero since the gravitational potential satisfies the Laplace's equation ($\Delta W = 0$).

In general, the gravity gradient tensor is measured with a gravity gradiometry (e.g., Lee, 2001; Barnes and Lumley, 2011; Dransfield and Christensen, 2013). However, since no gravity gradient surveys have been undertaken in Osaka yet, the gravity gradient tensor in this region was estimated using a technique based on the study of Mickus and Hinojosa (2001). The method described therein estimates a gravity gradient tensor as follows: (1) applying a Fourier transformation of the gravity anomaly, (2) estimating the gravitational potential by integration of the gravity anomaly in the Fourier domain, (3) calculating the gravity gradient components from second-order derivatives of the potential in each direction, and (4) applying Fourier inverse transformation to finally obtain all components of the tensor in the spatial domain.

In Figure 3, we show each component of the gravitational vector and gravity gradient tensor obtained from the residual Bouguer gravity anomaly shown in Figure 2 by Mickus and Hinojosa's method (Mickus and Hinojosa, 2001). Each component of the gravitational vector is estimated from the first-order derivatives of the potential in each direction. From Figure 3, we note that short wavelength signals are emphasized in each component of the gravity gradient tensor since they are derived as differential coefficients of the gravitational vectors.

2. Edge Emphasis

2.1. Various Edge Emphasis Techniques

Edge emphasis techniques are extraction techniques (e.g., Nabighian, 1972; Cordell and McCafferty, 1989; Cooper, 2012) of the steep horizontal gravity gradient belts caused by density structure changes.

In these techniques, the horizontal and vertical gravity gradient methods have been employed frequently for finding structure boundaries such as fault or contact between different materials (e.g., Blakely and Simpson, 1986; Shichi et al., 1992; Kudo and Kono, 1999; Yamamoto, 2003). The horizontal gravity gradient (*HG*) is given using the components of the gravity gradient tensor shown in equation (2) as

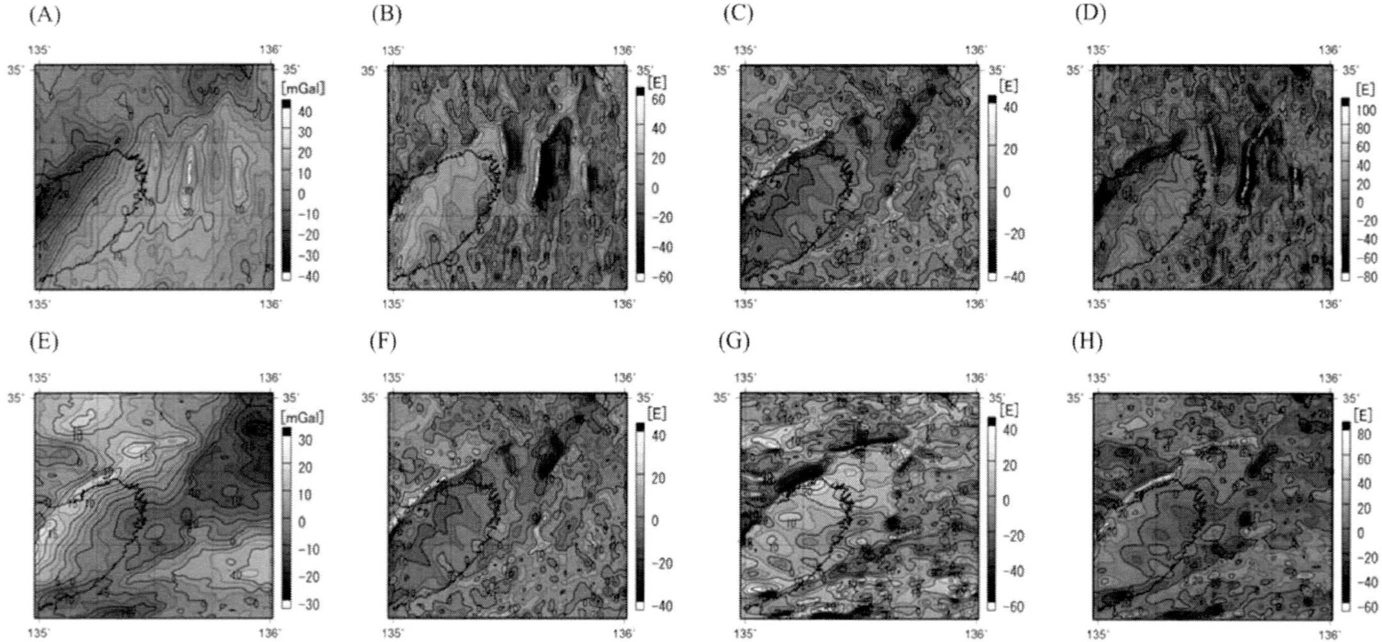

Figure 3. (Continued)

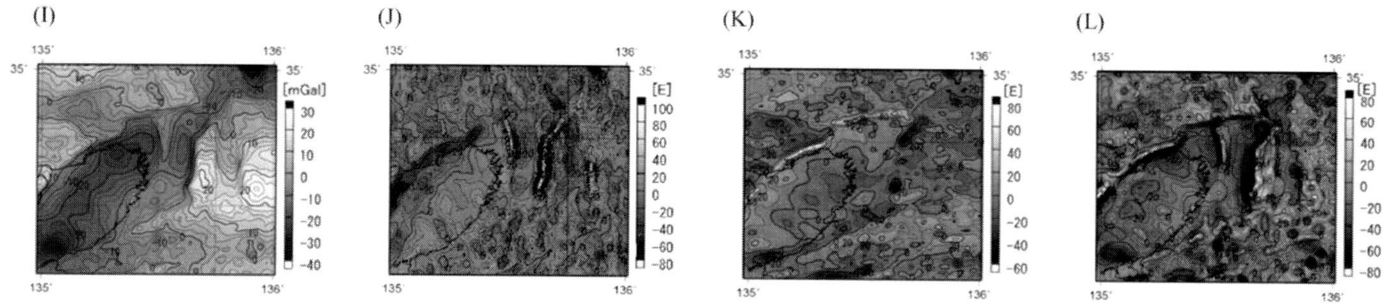

Figure 3. All components of gravitational vector and gravity gradient tensor. (A) g_x. Gravitational vector component in x-direction. (B) g_{xx}. Differential coefficient of g_x in the x-direction. (C) g_{xy}. Differential coefficient of g_x in the y-direction. (D) g_{xz}. Differential coefficient of g_x in the z-direction. (E) g_y. Gravitational vector component in y-direction. (F) g_{yx}. Differential coefficient of g_y in the x-direction. Component g_{yx} is equal to g_{xy}. (G) g_{yy}. Differential coefficient of g_y in the y-direction. (H) g_{yz}. Differential coefficient of g_y in the z-direction. (I) g_z. Gravitational vector component in z-direction. Gravity anomaly, well-known. (J) g_{zx}. Differential coefficient of g_z in the x-direction. Component g_{zx} is equal to g_{xz}. (K) g_{zy}. Differential coefficient of g_z in the y-direction. Component g_{zy} is equal to g_{yz}. (L) g_{zz}. Differential coefficient of g_z in the z-direction.

$$HG = \sqrt{g_{zx}^2 + g_{zy}^2} \tag{3}$$

Miller and Singh (1994) proposed the *TDR* (Tilt Derivative) defined by the arctangent of the ratio of the vertical gravity gradient to the *HG*.

$$TDR = \arctan\left(\frac{g_{zz}}{\sqrt{g_{zx}^2 + g_{zy}^2}} \right) \tag{4}$$

This emphasis technique does not extract extremely large amplitude signals in the short wavelength range, and is known as a balanced method. However, it is pointed out that the *TDR* extracts a phantom boundary at the zero line if positive and negative anomalies exist side by side (e.g., Cooper and Cowan, 2006; Ma, 2013).

Wijns et al. (2005) proposed the *THETA* map that could be used to obtain the structural boundaries,

$$THETA = \arccos\left(\frac{\sqrt{g_{zx}^2 + g_{zy}^2}}{\sqrt{g_{zx}^2 + g_{zy}^2 + g_{zz}^2}} \right) \tag{5}$$

and Cooper and Cowan (2006) proposed the *TDX* defined by the following equation.

$$TDX = \arctan\left(\frac{\sqrt{g_{zx}^2 + g_{zy}^2}}{|g_{zz}|} \right) \tag{6}$$

Li et al. (2014) conducted a numerical test of these emphasis techniques for structure boundaries extraction. As a result, they found that (1) *HG* method can find the structure boundaries of the causative body but its ability decreases with depth, (2) *TDR* and *THETA* methods are also able to retrieve the structure boundaries but their extracted shapes are not clear and, as mentioned above, these methods introduce a phantom boundary at the zero line in case of side by side positive and negative anomalies, and (3) *TDX* has the same phantom boundary limitation but the method can give clearer boundaries location than *TDR* and *THETA*.

2.2. Applications and Discussion

Here, we applied the *HG* and *TDX* to gravity anomalies and gravity gradient tensor of the study area, and discussed results obtained by these methods.

Figure 4 is a horizontal gradient (*HG*) map showing the area higher than 20 E. Here, E (Eötvös) is the unit of gravity gradient and 1E = 0.1 mGal/km. From the western side of the study area, we can see that the high horizontal gradient belts, corresponding to the locations of the Rokko-Awaji Fault System, Arima-Takatsuki Tectonic Line, Uemachi Fault, Ikoma Fault System and Nara-Tohen Fault, are emphasized. Although these steep horizontal gravity gradients can be also found in the Bouguer gravity anomaly map (Figure 2), the *HG* map represents them more clearly, and we can evaluate their dimension quantitatively. In this area, the Ikoma Fault System has the highest horizontal gradient value exceeding 95 E. The second highest is the Rokko-Awaji Fault System reaching 70 E at the southern part.

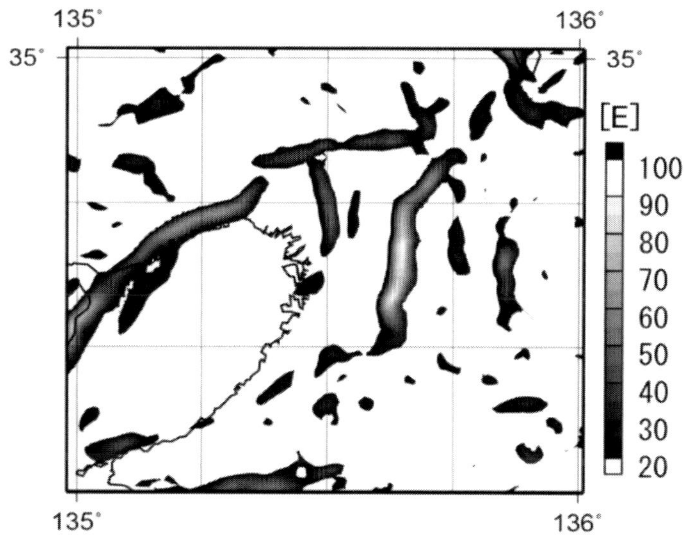

Figure 4. Horizontal gravity gradient map for values over 20 E. The contour interval is 5 E. E is the unit describing the gravity gradient, and 1 E = 0.1 mGal/km. mGal is the unit of gravitational acceleration, and mGal = 10^{-5} m/s^2.

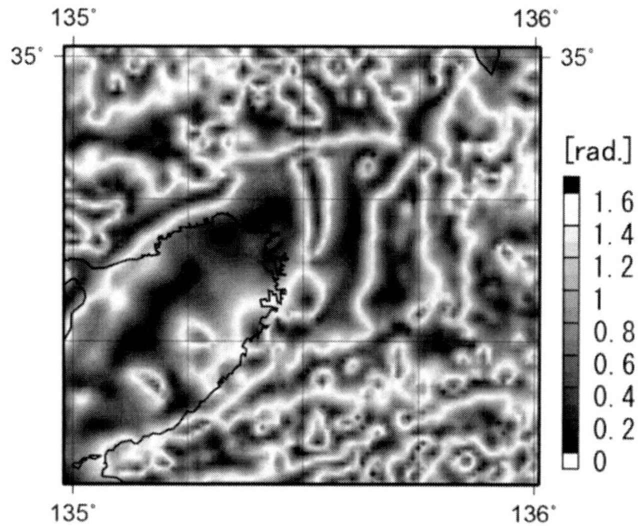

Figure 5. *TDX* map for emphasizing structural boundaries. The contour interval is 0.1 radian. The white lines of continuous maximum value are regarded as structural boundaries.

Figure 5 is the *TDX* map where lines of highest value (white) indicate the structure boundaries. Although we must beware the limitations of this method (extraction of a phantom boundary at the location of positive and negative anomalies juxtaposition), this map extracts tectonic lines, faults, and each outlines of sedimentary basins, the Uemachi highland, and the circular structures pointed out by Itoh et al. (2013). Especially, circular structures with strong magnetic anomalies in the Osaka Plain are clearly emphasized and we can find the same two structures in the Osaka Bay.

3. Euler Deconvolution

3.1. Basic Theories

Euler deconvolution is one of the semi-automatic interpretation methods based on Euler's homogeneity equation and is often employed to estimate locations and/or outlines of causative bodies. Because the Euler deconvolution technique can help quick interpretations of any potential field data in terms of depth and geological structures, it has been used by several researchers for analyzing both magnetic anomalies (e.g., Thompson, 1982; Barongo, 1984; Reid et al. 1990) and gravity anomalies (e.g., Marson and Klingele, 1993).

The Euler deconvolution based on the three orthogonal gradient components of the potential is called simply Euler deconvolution or Conventional Euler deconvolution. For the gravity anomaly g_z, it uses the equation

$$(x-x_0)g_{zx}+(y-y_0)g_{zy}+(z-z_0)g_{zz}=N(B_z-g_z) \quad (7)$$

In equation (7), x_0, y_0, and z_0 are unknown location parameters of the causative body center or of its edge that have to be estimated and x, y, and z are known location parameters of the observation point of the gravity and its gradients; N is the structural index; and B_z is the regional component of the gravity anomaly that has to be estimated.

Rewriting equation (7), we obtain the following equation and we can separate unknown parameters from known parameters:

$$x_0g_{zx}+y_0g_{zy}+z_0g_{zz}+NB_z=xg_{zx}+yg_{zy}+zg_{zz}+Ng_z \quad (8)$$

There are four unknown parameters in equation (8). If there were enough data, namely $n \geq 4$, to solve for these unknowns within a selected window, we could estimate these four parameters with the least squares method.

Zhang et al. (2000) suggested the Tensor Euler deconvolution, designed to consider the full gravity gradient tensor (equation (2)) and all components (g_x, g_y, g_z) of the gravitational vector. The Tensor Euler deconvolution is defined by the following simultaneous equations:

$$\begin{aligned}(x-x_0)g_{xx}+(y-y_0)g_{xy}+(z-z_0)g_{xz}&=N(B_x-g_x)\\(x-x_0)g_{yx}+(y-y_0)g_{yy}+(z-z_0)g_{yz}&=N(B_y-g_y)\\(x-x_0)g_{zx}+(y-y_0)g_{zy}+(z-z_0)g_{zz}&=N(B_z-g_z)\end{aligned} \quad (9)$$

Here, B_x, and B_y are the regional component of the gravity anomaly that have to be estimated. Thus, there are six unknown parameters in this equation, and if there were enough data to solve for these six unknowns, namely $n \geq 6$, within a selected window, these parameters could be estimated with the least squares method.

From a numerical test and an application to field data, Zhang et al. (2000) showed that the Tensor Euler deconvolution produces significantly improved resolution over the conventional Euler deconvolution.

3.2. Applications and Discussion

We employed here the Tensor Euler deconvolution (Zhang et al., 2000) to obtain information on subsurface structures. In the calculation, we set a 14 km width window (the required amount of data to solve equation (9) is 196) with a Structure Index of 0.5. The value of 0.5 for Structure Index is mostly used for the fine dike model (e.g., Rodrigues et al., 2014), but has also been employed for other general models such as thickness of the crust (e.g., Tedla, et al., 2011), and we found, after trial and error, that this value was the best suitable value.

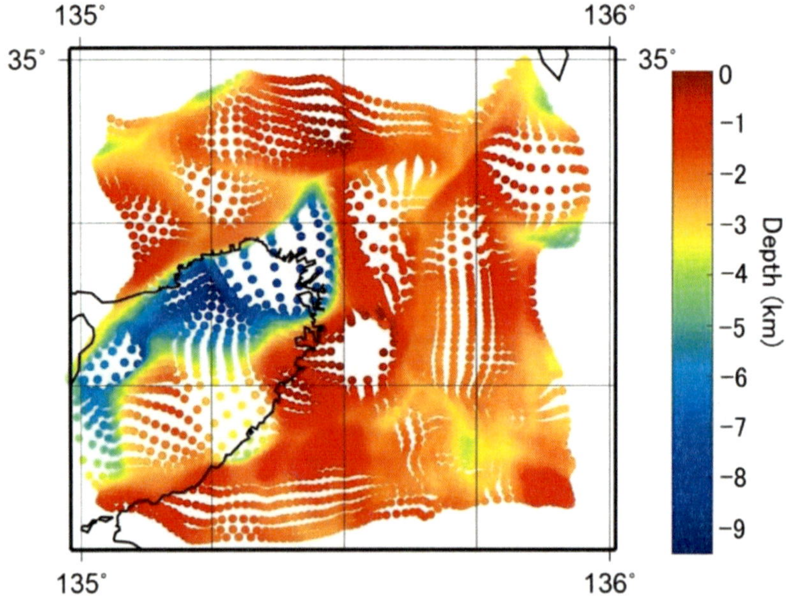

Figure 6. Solution clouds given by the Tensor Euler deconvolution. In the calculation, a 14km width window and a Structure index of 0.5 are assumed.

In Figure 6, we show the solution clouds given by the Tensor Euler deconvolution. If we consider the solution clouds representing roughly the basement structure in this region, the solutions give us interesting information on subsurface structures. From this figure, we find that the northern Osaka Bay is deeper than its southern part, and that the Rokko-Awaji Fault System and the Uemachi Fault have a steep dip angle toward the Osaka Bay area. In the Ikoma Fault System area, the basement is deeper toward the inland side. In addition, an outline of the circular structures in the Osaka Plain is extracted and presents a very shallow depth (0 – 1 km).

4. Dip Angle Estimation of Fault or Structural Boundary

4.1. Basic Theories

In the recent years, a technique estimating the dip angle of structures such as dike, fault or geological structure boundary has been developed as one of the semi-automatic interpretation methods using the gravity gradient tensor and has obtained good results.

Pedersen and Rasmussen (1990) defined the invariant ratio, I, of the gravity gradient tensor as follows

$$I = \frac{-27 I_2^2}{4 I_1^3} \tag{10}$$

Here, I_1 and I_2 are invariants of the tensor. Each invariant is given by three eigenvalues (λ_1, λ_2, λ_3) of the gravity gradient tensors as follows:

$$\begin{aligned} I_1 &= \lambda_1\lambda_2 + \lambda_2\lambda_3 + \lambda_1\lambda_3 \\ I_2 &= \lambda_1\lambda_2\lambda_3 \end{aligned} \tag{11}$$

The invariant ratio I is distributed between 0 and +1. If the body causing a gravity anomaly is a 2-D structure, I is 0. If the causative body is 3-D, I is +1. Beiki and Pedersen (2010) named this ratio the Dimensionality Index, and suggested I = 0.5 to be the threshold between a 2-D and 3D causative structure. In addition, they pointed out that, in case of a 2-D causative structure such as a dike, the maximum eigenvector of the gravity gradient tensor points to the causative body and the minimum eigenvector is parallel to the strike direction of the structure (Figure 7).

Beiki (2013) suggested that the dip angle, α, of the causative body could be estimated from the three components (x, y, z) of the maximum eigenvector (v_1) of the gradient tensor of the potential field as (Figure 7):

$$\alpha = \arctan\left(\frac{v_{1z}}{\sqrt{(v_{1x})^2 + (v_{1y})^2}}\right) \tag{12}$$

where v_{1x}, v_{1y} and v_{1z} are the x, y, and z components of the maximum eigenvector, v_1. Beiki (2013) applied this method to an aeromagnetic data set of the Åsele area in Sweden, and provided useful information on the dip angle of dike swarms. Kusumoto (2015) applied this technique to estimate the dip angle of the Median Tectonic Line located in the southwest Japan, and obtained a dip angle distribution compatible with the results from other geophysical surveys such as reflection survey.

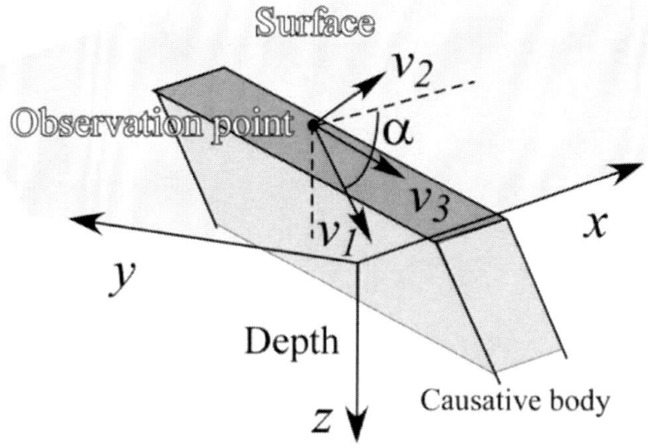

Figure 7. Schematic illustration of the eigenvectors for a 2-D structure. In this figure, v_1, v_2, and v_3 are the maximum, intermediate, and minimum eigenvectors, respectively. The angle α between the surface and maximum eigenvector is the angle of the structural boundary and/or fault.

4.2. Applications and Discussion

Figure 8 is the Dimensionality Index map where areas of $I \leq 0.5$ are gray-colored and characterized by 2-D structures. These 2-D structures are distributed widely in the study area, which has many active faults.

Consequently, we reasoned that faults and/or long narrow structures caused by fault motions cover the majority of this area. As the steep horizontal gravity gradient belts correspond very well with the active fault traces, we next attempted to estimate the dip angle of each active fault using the eigenvectors of the gravity gradient tensor.

In this study, the dip angle of the structure was estimated for cases where the Dimensionality Index was lower than 0.5 and the horizontal gravity gradient was higher than 20 E. Results are shown in Figure 9.

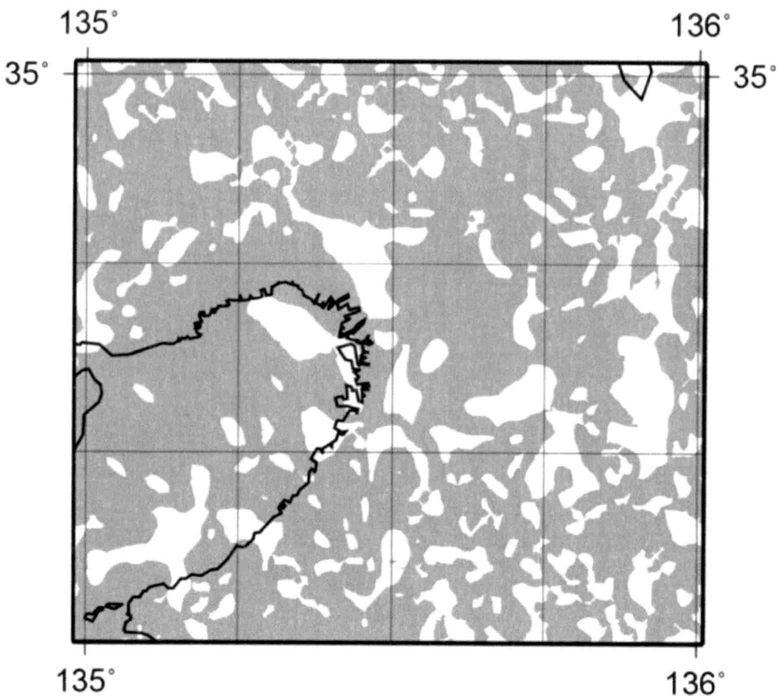

Figure 8. Map of dimensionality index, I. Colored areas indicate $I < 0.5$, where 2-D structures can be predicted.

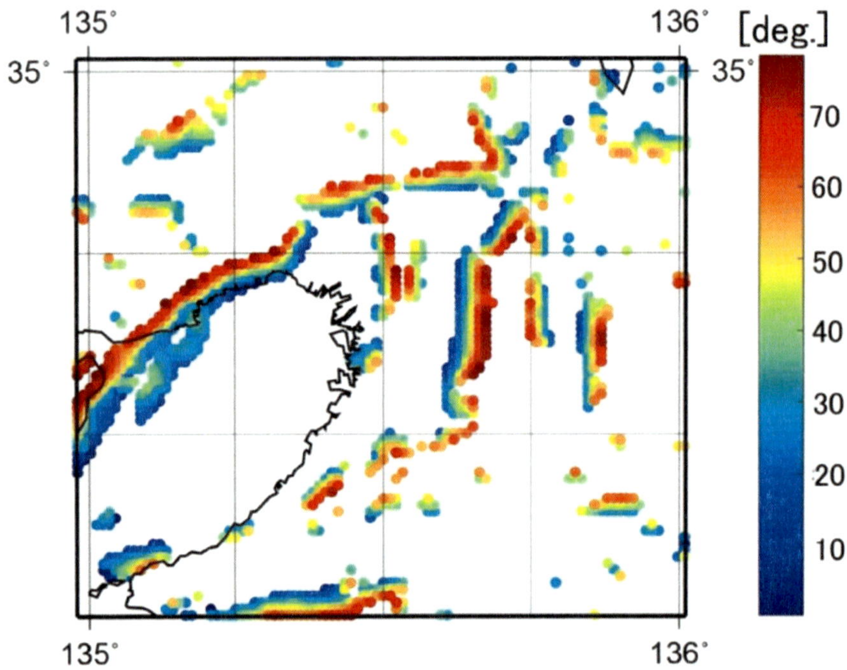

Figure 9. Distribution of estimated dip angles of faults and/or structural boundaries. Dip angle estimation was conducted for the areas that satisfied the conditions $I < 0.5$ and HG (Horizontal gravity gradient) ≥ 20 E.

We found that most of the faults, fault systems and tectonic lines in this area have dip angles higher than 55 degrees. The Uemachi fault and the Arima-Takatsuki Tectonic line have dip angles in the range of 55-65 degrees. The Ikoma Fault System and the Rokko-Awaji Fault System have high dip angles reaching 70 degrees or even exceeding 70 degrees in some parts. The Ikoma Fault System has high dip angles, particularly on the inland side. These values are supported by the results obtained with the Tensor Euler deconvolution (Figure 6). Since most of the active faults in this area have both lateral and reverse components, these fault dip angle estimations are realistic and reasonable.

Conclusion

We calculated the gravitational vector and gravity gradient tensor from the Bouguer gravity anomaly, and we applied them to location estimations of causative bodies and to their dip angle estimations.

As a result, characteristic anomalies such as active faults and circular structures with strong magnetic anomalies were extracted by edge emphasis techniques. In addition, it was found that the Rokko-Awaji Fault System and the Uemachi Fault have a steep dip angle toward the Osaka Bay area, and that these active faults have dip angles in the range of 55-65 degrees (exceeding 70 degrees in some parts).

Since most of the active faults in this area have both lateral and reverse components, these fault dip angle estimations are realistic and reasonable.

Acknowledgments

This work was supported partially by JSPS (Japan Society for the Promotion of Science) KAKENHI Grant Number 15K14274.

References

Barnes, G., & Lumley, J. (2011). Processing gravity gradient data. *Geophysics,* 76, I33–I47.

Barongo, J. O. (1984). Euler's differential equation and the identification of the magnetic point-pole and point-dipole sources. *Geophysics,* 49, 1549-1553.

Beiki, M., & Pedersen, L. B. (2010). Eigenvector analysis of gravity gradient tensor to locate geologic bodies. *Geophysics,* 75, I37–I49. doi:10.1190/1.3484098.

Beiki, M. (2013). TSVD analysis of Euler deconvolution to improve estimating magnetic source parameters: an example from the Asele area, Sweden. *Journal of Applied Geophysics,* 90, 82-91. doi:10.1016/ j.jappgeo2013.01.002.

Blakely, R., & Simpson, R. W. (1986). Approximating edges of source bodies from magnetic or gravity anomalies. *Geophysics,* 51, 1494-1498.

Cordell, L., & McCafferty, A. E. (1989). A terracing operator for physical property mapping with potential field data. Geophysics, 54, 621-634.

Cooper, G. R. J., & Cowan, D. R. (2006). Enhancing potential field data using filters based on the local phase. *Computers and Geosciences,* 32, 1585-1591.

Cooper, G. R. J. (2012). The removal of unwanted edge contours from gravity datasets. *Exploration Geophysics,* 44, 42-47.

Dransfield, M. H., & Christensen, A. N. (2013). Performance of airborne gravity gradiometers. *The Leading Edge,* 32, 908-922. doi:10.1190/tle32080908.1.

Hofmann-wellenhof, B., & Moritz, H. (2005). *Physical Geodesy.* Berlin: Springer.

Itoh, Y., Kusumoto, S., & Takemura, K. (2013). Characteristic basin formation at terminations of a large transcurrent fault - basin configuration based on gravity and geomagnetic data. In Y. Itoh (Ed.), Mechanism of Sedimentary Basin Formation - *Multidisciplinary Approach on Active Plate Margins.* Rijeka (Croatia): InTech. http://dx.doi.org/10.5772/56702.

Kudo, T., & Kono, Y. (1999). *Relationship between distributions of shallow earthquakes and gradient of gravity anomaly field in southwest Japan.* Zishin 2, 52, 341-350. (in Japanese with English abstract)

Kusumoto, S., Fukuda, Y., Takemoto, S., & Yusa, Y. (1996). Three-dimensional subsurface structure in the eastern part of the Beppu-Shimabara Graben, Kyushu, Japan, as revealed by gravimetric data. *Journal of Geodetic Society of Japan,* 42, 167-181.

Kusumoto, S. (2015). Estimation of dip angle of fault or structural boundary by eigenvectors of gravity gradient tensors. *Butsuri-Tansa,* 68, 277-287. (in Japanese with English abstract)

Lee, J. B. (2001). Falcon gravity gradiometer technology. *Exploration Geophysics,* 32, 247-250.

Li, L., Huang, D., Han, L., & Ma, G. (2014). Optimized detection filters in the interpretation of potential field data. *Exploration Geophysics,* doi: 10.1071/EG13059.

Ma, G. (2013). Edge detection of potential field data using improved local phase filter. *Exploration Geophysics,* 44, 36-41.

Mickus, K. L., & Hinojosa, J. H. (2001). The complete gravity gradient tensor derived from the vertical component of gravity: a Fourier transform technique. *Journal of Applied Geophysics,* 46, 159-174.

Miller, H. G., & Singh, V. (1994). Potential field tilt – A new concept for location of potential field sources. *Journal of Applied Geophysics,* 32, 213-217.

Marson, I., & Klingele, E. E. (1993). Advantages of using the vertical gradient of gravity for 3-D interpretation. *Geophysics,* 58, 1588-1595.

Nabighian, M. N. (1972). The analytic signal of two-dimensional magnetic bodies with polygonal cross-section: its properties and use for automated anomaly interpretation. *Geophysics,* 37, 507-517.

Perdersen, L. B., & Rasmussen, T. M. (1990). The gradient tensor of potential field anomalies: some implications on data collection and data processing of maps. *Geophysics,* 55, 1558-1566.

Reid, A. B., Allsop, J. M., Granser, H., Millett, A. J., & Somerton, I. W. (1990). Magnetic interpretation in three dimensions using Euler deconvolution. *Geophysics,* 55, 80-91.

Rodrigues, R. S., de Castro, D. L., & dos Reis, J. A. Jr. (2014). Characterization of the Potiguar rift structure based on Euler deconvolution. *Revista Brasileira de Geofisica,* 32, 109-121.

Shichi, R., Yamamoto, A., Kimura, A., & Aoki, H. (1992). Gravimetric evidences for active faults around Mt. Ontake, central Japan: specifically for the hidden faulting of the 1984 Western Nagano Prefecture Earthquake. *Journal of Physics of the Earth,* 40, 459-478.

Tedla, G. E., van der Meijde, M., Nyblade, A. A., & van der Meer, F. D. (2011). A crustal thickness map of Africa derived from a global gravity field model using Euler deconvolution. *Geophysical Journal International,* 187, 1-9.

ten Brink, U. S., Ben-Avarham, Z., Bell, R. E., Hassouneh, M., Coleman, D. F., Andreasen, F., Tibor, G., & Coakley, B. (1993). Structure of the Dead Sea pull-apart basin from gravity analyses. *Journal of Geophysical Research,* 98, 21877-21894.

Thompson, D. T. (1982). EULDPH – A new technique for making computer assisted depth estimates from magnetic data. *Geophysics,* 47, 31-37.

Torge, W. (1989). Gravimetry. Berlin: Walter de Gruyter.

Wijns, C., Perez, C., & Kowalezyk, P. (2005). Theta map: edge detection in magnetic data. *Geophysics,* 70, L39-L43.

Yamamoto, A. (2003). Gravity anomaly atlas of the Ishikari Plain and its vicinity, Hokkaido, Japan. *Geophysical Bulletin of Hokkaido,* 66, 33-62. (in Japanese with English abstract)

Zhang, C., Mushayandebvu, M. F., Reid, A. B., Fairhead, J. D., & Odegrad, M. E. (2000). Euler deconvolution of gravity tensor gradient data. *Geophysics,* 65, 512-520.

Chapter 3

SUBSURFACE STRUCTURE OF OSAKA SEDIMENTARY BASIN AND ITS TECTONIC EVOLUTION

Yasuto Itoh*

Graduate School of Science, Osaka Prefecture University, Osaka, Japan

ABSTRACT

The three-dimensional architecture of the Osaka sedimentary basin has been clarified based on seismic interpretation. The tectonic depression is divided into northern and southern sectors. The former develops horst and graben structures under prevailing east–west compressive stress, whereas the latter shows divergent features reflecting the temporal transition of tectonic regimes. A timeline of tectonic episodes incorporating new radiometric data of fluvial sediments is discussed in this chapter.

1. INTRODUCTION

As a result of the heavy industrialization since the early 20th century, Osaka has become a site of serious land subsidence, and a number of stratigraphic test wells have been planned to determine the physical properties

* Correspondence: yasutokov@yahoo.co.jp.

of the basin-filling Quaternary clastics. Continuous sediment records have provided stratigraphic standards for the Middle and Upper Subgroups of the Osaka Group (e.g., Ikebe et al., 1970) and led to the development of basin evolution models linked to regional tectonics (e.g., Huzita, 1980). A sudden advancement in geophysical research was provoked by the 1995 Hyogoken–Nanbu Earthquake (M7.2), which devastated the urban area of Kobe. With the aim of preventing impending disasters, many seismic surveys have since been conducted to visualize the three-dimensional architecture of the basin and evaluate the numerous active faults running through the unconsolidated strata. Another technological improvement was brought about by the improvement of U–Pb dating utilizing inductively coupled plasma mass spectrometry (ICP-MS) combined with an excimer laser ablation (LA) sample introduction system, which greatly improved the accuracy of the chronological discussion. Based on these enhancements to the spatiotemporal resolution of available datasets, this chapter aims to clarify the evolutionary process of the Osaka sedimentary basin.

2. Seismic Interpretation

In an effort to comprehend the three-dimensional architecture of the Osaka sedimentary basin, Itoh et al. (2013a, 2015) published geological interpretations of seismic profiles in the tectonic basin. This article presents several unpublished datasets, which were chosen to highlight the structural diversity of the basin. From north to south, basic specifications of the seismic lines are described in the following paragraphs (see Figure 1 for line locations).

In 1996, the National Research Institute for Earth Science and Disaster Prevention conducted a seismic survey of the approximately 8.5-km-long Maishima No. 1 line on the northern margin of the Osaka sedimentary basin (Figure 1; Kasahara et al., 1997). During the shooting, 240 channels of geophones at 25-m intervals recorded the energy released from two vibrators shot at 50-m intervals. The raw seismic data were stacked and then subjected to a post-stack processing sequence to enhance the resolution.

In 1995, Osaka City conducted a seismic survey (Osaka City, 1995). During the shooting of the approximately 11-km-long Osaka No. 1 line in the central part of the Osaka sedimentary basin (Figure 1), 120 channels of geophones at 20-m intervals recorded the energy released from one vibrator

(Y2400) shot at 20-m intervals. The raw seismic data were stacked and then subjected to a post-stack processing sequence to enhance the resolution.

In 1996, the Geological Survey of Japan conducted a seismic survey (Geological Survey of Japan, 1997). During the shooting of the approximately 3-km-long GS Yao line on the western foot of the Ikoma Mountains (Figure 1), 60 channels of geophones at 10-m intervals recorded the energy released from one Yuatsu Impactor (non-explosive land seismic energy source; JMI-200) shot at 5-m intervals. The raw seismic data were stacked and then subjected to a post-stack processing sequence to enhance the resolution.

In 2004, Osaka Prefecture conducted a seismic survey (Osaka Prefecture, 2005). During the shooting of the approximately 6-km-long Yao line on the western foot of the Ikoma Mountains (Figure 1), 126 channels of geophones at 20-m intervals recorded the energy released from one vibrator (Y2400) shot at 20-m intervals. The raw seismic data were stacked and then subjected to a post-stack processing sequence to enhance the resolution.

In 2003, Osaka Prefecture conducted another seismic survey (Osaka Prefecture, 2004). During the shooting of the approximately 9-km-long Wangan (South) line along the Osaka Bay coastline (Figure 1), 126 channels of geophones at 20-m intervals recorded the energy released from one vibrator (Y2400) shot at 20-m intervals. Raw seismic data were stacked and then subjected to a post-stack processing sequence to enhance the resolution. This line crosses the coast-normal interpreted lines described below.

In 2011, the Disaster Prevention Research Institute of Kyoto University conducted a seismic survey (Research and Development Bureau, MEXT and DPRI, Kyoto University, 2012). During the shooting of the approximately 11-km-long Otsugawa line on the Osaka Bay coast (Figure 1), 150 channels of geophones at 20-m intervals recorded the energy released from one vibrator (EnviroVibe) shot at 10-m intervals. The raw seismic data were stacked and then subjected to a post-stack processing sequence to enhance the resolution.

In 2012, the Disaster Prevention Research Institute of Kyoto University conducted another seismic survey (Research and Development Bureau, MEXT and DPRI, Kyoto University, 2012). During the shooting of the approximately 10-km-long Takaishi–Sakai line on the Osaka Bay coast (Figure 1), 100 channels of geophones at 20-m intervals recorded the energy released from one vibrator (Hemi-40) shot at 10-m intervals. The raw seismic data were stacked and then subjected to a post-stack processing sequence to enhance the resolution.

2.1. Northern Sector

Surrounded by remarkably active faults, the northern sector of the Osaka basin has been in the compressive regime through the Quaternary. Figure 10 presents the seismic profiles of the GS Yao and Yao lines on the western foot of the Ikoma Mountains (Figure 1). It is obvious that the steep gradient of the Bouguer anomaly was caused by a group of reverse faults, collectively named the Ikoma Fault, the separation of which reaches approximately 500 m based on the seismic horizon of the erosional surface of the acoustic basement. The two-dimensional gravity modeling of the fault showed that the cumulative vertical motion exceeds 1500 m (Horike et al., 1995). It should be noted that the deposits of the ancient Yamato River (Miyachi et al., 1998), which was dried up after the excavation of an east–west canal in AD 1704, are distributed along geomorphic minima. Itoh et al. (2015) demonstrated that the natural river path coincides with a subsidence area indicated by a dislocation model assuming the reverse motion on the Ikoma Fault.

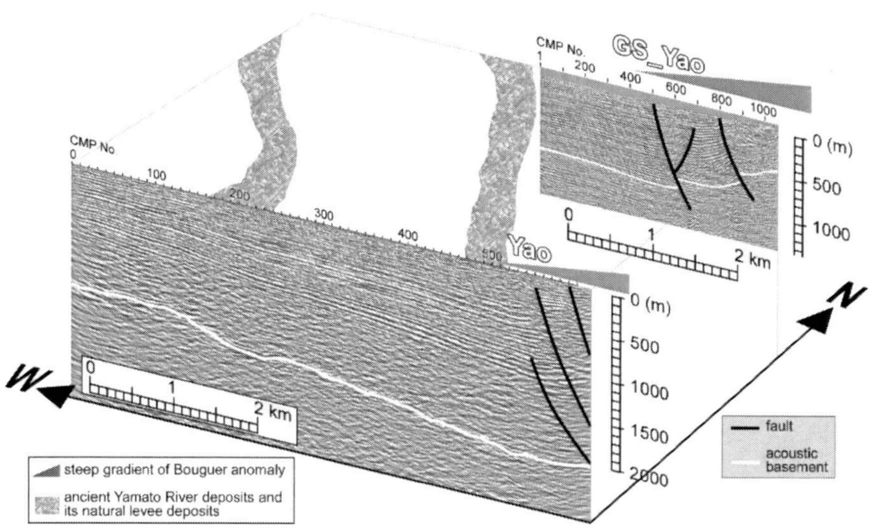

Figure 10. Depth-converted seismic profiles without vertical exaggeration around the western foot of the Ikoma Mountains. See Figure 1 for line locations.

Figure 11 shows a depth section of the Maishima No. 1 line on the northern margin of the Osaka sedimentary basin (Figure 1). The uplifted area in the eastern part is an extension of a remarkable north–south gravitational trend (Figure 1), the surface flexure of which is named the Uemachi Fault.

Horike et al. (1996) determined the S-wave velocities in the basin and estimated the vertical separation on the fault to be approximately 500 m. In this seismic profile, the large reverse fault on the western side of the horst (Figure 11; CMP No. 400) was defined as the Butsunenjiyama Fault, which deforms the Quaternary Osaka Group (Huzita and Kasama, 1982). The horst and graben features may express contractional deformation, and this study considers the coexistence of normal and reverse faulting in a series of structures. It is characteristic of collapsed anticlines known as a specific deformation along strike-slip faults (e.g., Itoh et al., 2014). Because the Osaka basin is bordered by the dextral Arima–Takatsuki Tectonic Line on its northern end (Huzita, 1980; Huzita and Kasama, 1982), the simple shear mode was considered in the interpretation of the Maishima No. 1 line.

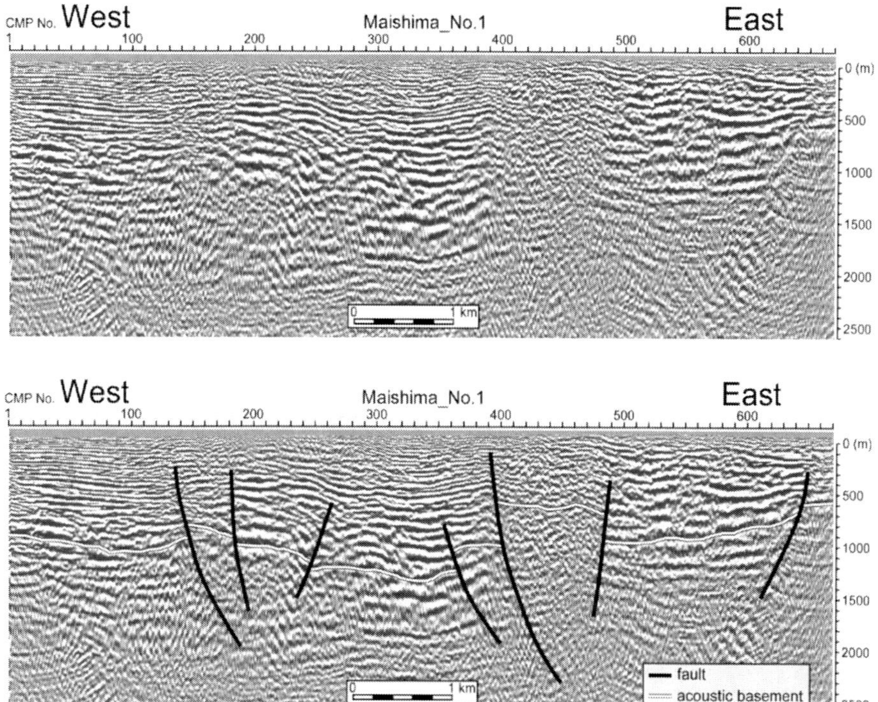

Figure 11. Depth-converted seismic profiles (top: raw, bottom: interpreted) of Maishima No. 1 line without vertical exaggeration. See Figure 1 for line location.

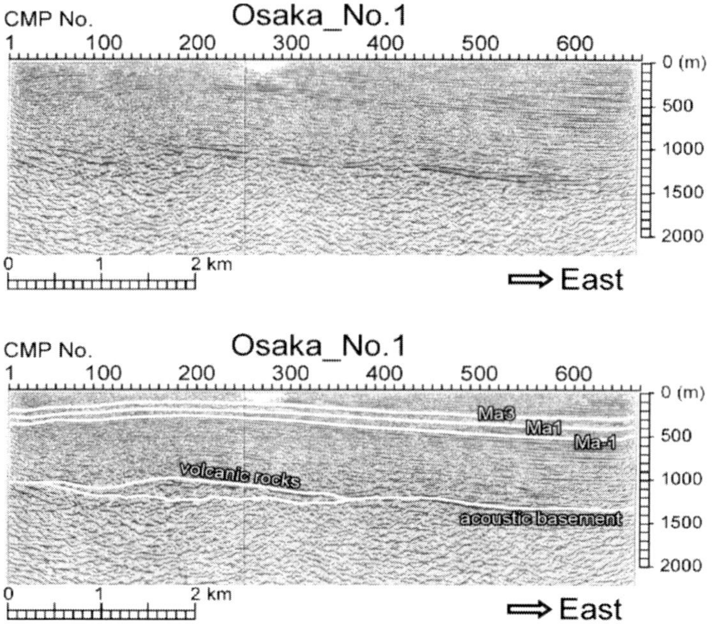

Figure 12. Depth-converted seismic profiles (top: raw, bottom: interpreted) of Osaka No. 1 line without vertical exaggeration. See Figure 1 for line location.

Figure 12 shows a depth section of the Osaka No. 1 line set in the midst of the densely populated Osaka metropolis. The weakened gravitational gradient (Figure 1) indicates that there is no clear displacement in the basin-filling strata, as stated in a survey summary by Kondo et al. (2015). Although they regarded the rugged surface of the acoustic basement to have been developed by faulting, their interpretation was founded on the unrealistic assumption of a two-layer model comprising granitic basement and unconsolidated sediments. Actually, strong discontinuous reflectors in the deep interior of the basin are accompanied by positive geomagnetic anomalies (Figure 1), implying volcanic activities prior to the Quaternary basin formation. For a domal structure on the Yamatogawa South line (Figure 1), Itoh et al. (2013a) performed three-dimensional Talwani modeling on the assumption that a concealed conical volcano settled during an extensive volcano-tectonic event in the Miocene and successfully restored the observed gravity and geomagnetic anomaly. Thus, the heterogeneous nature of the upper crust constituents should be taken into consideration when reconstructing the evolutionary process of the Osaka sedimentary basin.

2.2. Southern Sector

Although the present Yamato River is an artificial canal excavated in the Edo era, as mentioned above, it has unexpectedly acted as a border between the northern and southern sectors of the studied basin. Namely, the southern side of the river is characterized by a multi-azimuthal gravity anomaly trend and a sporadic positive geomagnetic anomaly (see Figure 1). The former implies the occurrence of temporal changes in stress regimes, and the latter may be related to voluminous buried volcanic rocks.

Figure 13 shows a depth section of the Otsugawa line on the Osaka Bay coast (Figure 1). A minor reverse fault around the southeastern region of the profile (CMP Nos. 1700–1800) corresponds to the active trace of the Kumedaike Fault (Okada and Togo, 2000). Strong deep reflectors on the downthrown side of the fault have been interpreted as a volcanic surface based on the positive geomagnetic anomaly around the seismic line. It should be noted that the inclined fault blocks in the northwestern half suggest the local development of extensional features. The seismic horizons in this profile are correlated with those on the Takaishi–Sakai line (next paragraph) via the crossing Wangan line (Figure 1).

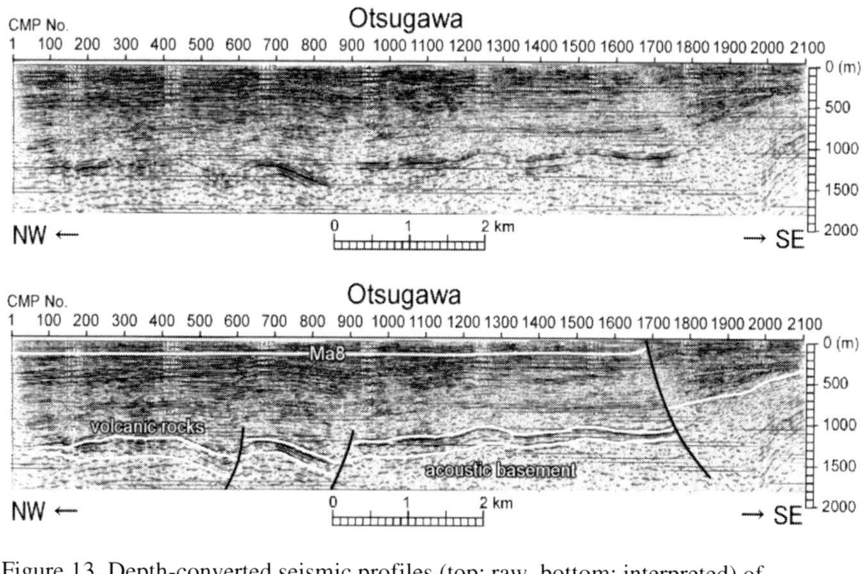

Figure 13. Depth-converted seismic profiles (top: raw, bottom: interpreted) of Otsugawa line without vertical exaggeration. See Figure 1 for line location.

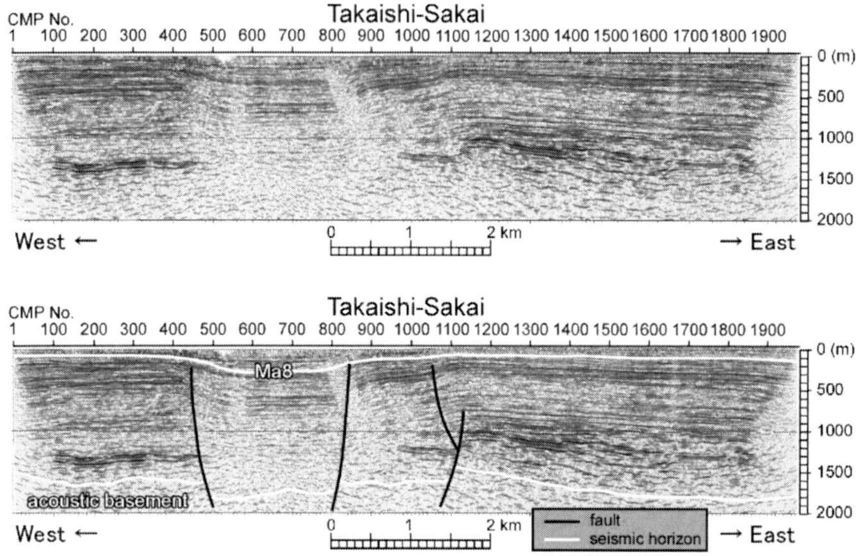

Figure 14. Depth-converted seismic profiles (top: raw, bottom: interpreted) of Takaishi–Sakai line without vertical exaggeration. See Figure 1 for line location.

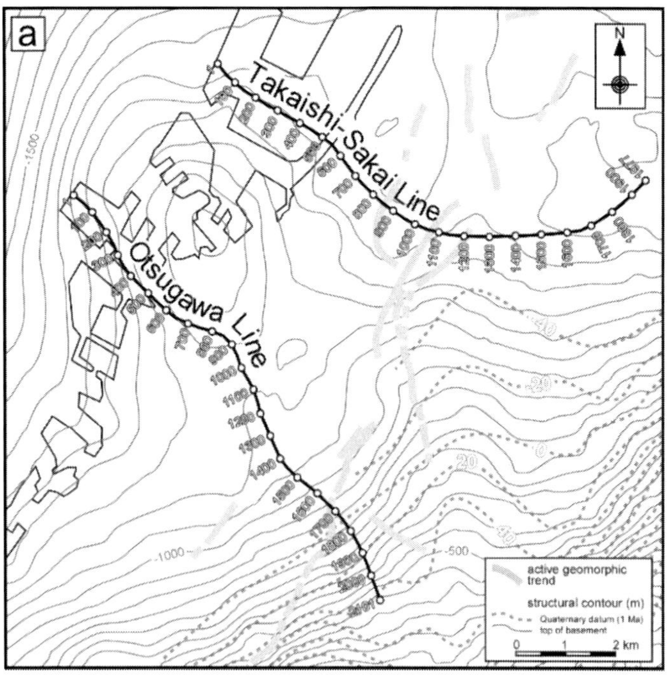

Figure 15. (Continued)

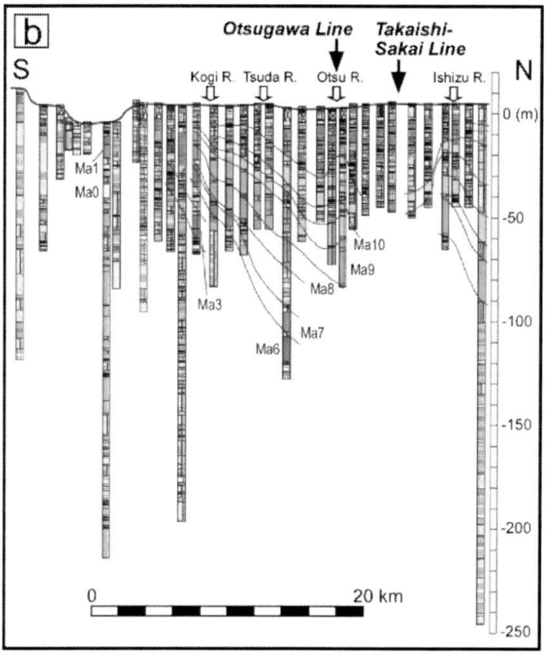

Figure 15. (a) Structural trend around survey areas of Otsugawa and Takaishi–Sakai lines. The active geomorphic trends (thick solid line) and the structure of the top of the acoustic basement (thin solid line) are after the Disaster Prevention Research Institute (2013). The structure delineated by a widespread tephra in the Quaternary Osaka Group (dashed line) is after Itihara et al. (1986). (b) Along-coast structure of upper part of Osaka Group based on boring database (Research Committee on Ground in Kansai, 2007).

Figure 14 shows a depth section of the Takaishi–Sakai line on the Osaka Bay coast (Figure 1). Minor superficial undulations aside, a notable feature of this profile is a tension graben around the coastline (CMP Nos. 500–800). The local extension is again indicated within the southern sector, and the tectonic context may be elucidated by compiling subsurface information.

Figure 15a shows structural trends around the survey areas of the Otsugawa and Takaishi–Sakai lines. The structure of the top of the acoustic basement (Disaster Prevention Research Institute, 2013) and that in the Osaka Group delineated by a widespread tephra at 1 Ma (Itihara et al., 1986) show close resemblance in the shapes of their north–south broad anticlines, which are surrounded by surface geomorphic trends. The above-mentioned sag on the Takaishi–Sakai line is located near the northern tip of the recent nose structure, where local tensile stress inevitably exists. The along-coast structure of the upper part of the Osaka Group based on a boring database (Figure 15b; Research Committee on Ground in Kansai, 2007) clearly demonstrates that the datum planes are at their lowest around the Takaishi–Sakai line and implies a bowl-like shape of neotectonic sag. Thus, the strange structure is interpreted in the framework of a larger structural trend that likely developed under an east–west compressive regime. One final important point is that the top of the acoustic basement and the Late Quaternary datum are similar in their morphology but the structural amplitude of the latter is one digit smaller than that of the former, which suggests the repeated buildup of the structure. According to the recent geomorphic surface trends obtained by Oka (1961), the development of the neotectonic feature appears to still be underway.

3. Chronology of Ancient Fluvial Basin

Seismic analysis has revealed conspicuous tectonic episodes during the Quaternary. As Itoh et al. (2013b) noted, the temporal transition of stress regimes was likely closely related to the change in the subduction mode of the Philippine Sea Plate. It is well known that a latitudinal fluvial basin developed in the late Quaternary at the southern foot of the Izumi Mountains, a fact indicative of a tectonically dormant interval. To elucidate the time sequence of such episodes, radiometric dating was performed on volcanic ash samples intercalated in the coarse clastics.

3.1. Samples

The southern sector of the Osaka sedimentary basin is bounded by the Izumi Mountains. Its uplift history has been controlled by slip-sense changes on the Median Tectonic Line (MTL) running along the southern foothill. As shown in Figure 16a, the MTL is a longstanding transcurrent fault representing a regional crustal break on the eastern Eurasian margin and was connected to the Central Sikhote Alin Fault around the K/T boundary. It was initially activated as a sinistral fault, which is reflected in the echelon arrangement of the metamorphic grade zonation of the Sanbagawa belt (Figure 16b; Wang and Maekawa, 1997). From the steep gradient of the gravity anomaly, deep-rooted activity on the MTL is conceivable. Hirota (1991) found a remarkable discontinuity in the metamorphic grade around the Funaokayama bar in the Kinokawa River (star in Figure 16b) and regarded it as a tectonic block.

Recent MTL trace reactivated since the Pliocene was found on the northern bank of the river, and dextral and/or reverse faulting was confirmed by the Research Group for Active Faults (1991). The downthrown side of the fault is intermittently buried by fluvial sediments of the Shobudani Group.

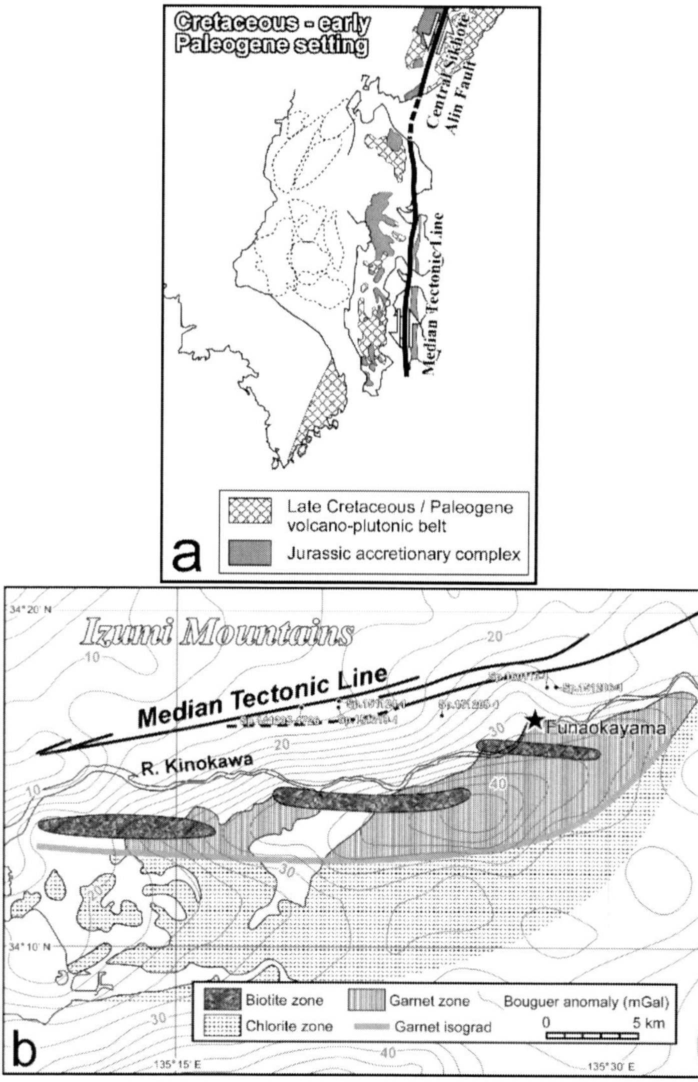

Figure 16. Index map of ash sampling sites for radiometric dating on the southern foothill of the Izumi Mountains, with tectonic context of the Median Tectonic Line.

Based on tephrochronology, Makimoto et al. (2004) defined the Shobudani Group as a series of unconsolidated fluviatile deposits corresponding to the marine–nonmarine sequence of the Osaka Group filling the target area of the present study. From radiometric dating and plant fossil assemblages, the Shobudani Group is divided into the Kishigawa, Kaseda, and Toge Formations in ascending order. Five analyzed ash samples (Sp. 141205-4726, 151124-1, 151219-1, 160116-1, and 151206-1) were obtained from the Kaseda Formation, and one (Sp. 151205-1) was obtained from the Toge Formation (see Figure 16b for locations).

3.2. Methods

The fission track (FT) dating method used in this study was developed by Danhara and Iwano (2009). U–Pb age data were obtained using ICP-MS combined with an excimer LA sample introduction system. The U–Pb ages of zircon samples were determined after chemical leaching using 47% HF for 20 h at room temperature or after fission track (FT) etching using a KOH–NaOH eutectic solution for 36, 43, or 45 h at 225°C.

3.3. Results

The results of the FT dating of zircon grains are summarized in Tables 1a–6a and Figures 17a–22a. The colorless euhedral crystals of Sp. 151124-1 (Figure 17a) were analyzed because it contained many varicolored grains derived from older terranes. Sp. 151205-1 (Figure 18a) contained numerous zircons of basement rocks, and colorless crystals were selected for measurement. Although 11 grains among a total of 30 had no spontaneous FT, their ages were tightly clustered and yielded a positive value in a chi-squared test. The single-grain FT ages of Sp. 151206-1 (Figure 19a) showed a large scatter, and the youngest concordant cluster in the U–Pb analysis was then adopted for the FT count. Although Sp. 151219-1 (Figure 20a) contained numerous euhedral monogenetic zircons, their mean age (Early Miocene) appears completely discordant with the stratigraphic framework, and the sampled layer is considered to be reworked sediments of older volcanic rocks. The single-grain FT ages of Sp. 160116-1 (Figure 21a) showed a large scatter, and the youngest concordant cluster in the U–Pb analysis was then adopted for the FT count. Sp. 141205-4726 (Figure 22a) contained a moderate amount of homogeneous zircon grains, and a reliable weighted mean age was obtained from this sample.

Table 1a. Fission-track grain ages of zircons obtained from Sp. 151124-1

No.	Ns	Nu	S $\times 10^{-5}$ (cm^{2})	ρ_s $\times 10^{-7}$ (cm^{-2})	ρ_u $\times 10^{-9}$ (cm^{-2})	Ns / Nu $\times 10^{3}$	T (Ma)	σ_T (Ma)	U (ppm)
22	3	3618	1.20	2.50	0.30	0.83	4.71	2.73	180
24	7	9893	2.80	2.50	0.35	0.71	4.02	1.53	211
14	1	1532	0.90	1.11	0.17	0.65	3.71	3.71	101
13	3	8367	2.70	1.11	0.31	0.36	2.04	1.18	185
2	17	50235	8.40	2.02	0.60	0.34	1.92	0.47	357
4	2	5949	4.00	0.50	0.15	0.34	1.91	1.35	89
30	11	40132	4.90	2.24	0.82	0.27	1.56	0.47	488
3	4	15467	4.20	0.95	0.37	0.26	1.47	0.74	220
9	2	9012	4.00	0.50	0.23	0.22	1.26	0.89	134
23	1	4853	2.00	0.50	0.24	0.21	1.17	1.17	145
19	2	9821	4.00	0.50	0.25	0.20	1.16	0.82	146
29	1	5586	2.40	0.42	0.23	0.18	1.02	1.02	139
15	2	11525	2.40	0.83	0.48	0.17	0.99	0.70	286
5	2	12420	5.00	0.40	0.25	0.16	0.92	0.65	148
1	3	20396	7.00	0.43	0.29	0.15	0.84	0.48	174
6	0	16150	5.40	0.00	0.30	0.00	0.00	****	178
7	0	2130	1.40	0.00	0.15	0.00	0.00	****	91
8	0	1776	1.50	0.00	0.12	0.00	0.00	****	71
10	0	2934	0.80	0.00	0.37	0.00	0.00	****	219
11	0	3194	1.20	0.00	0.27	0.00	0.00	****	159
12	0	3092	1.50	0.00	0.21	0.00	0.00	****	123
16	0	59202	4.00	0.00	1.48	0.00	0.00	****	882
17	0	3485	1.80	0.00	0.19	0.00	0.00	****	115
18	0	612	0.90	0.00	0.07	0.00	0.00	****	41
20	0	1583	1.00	0.00	0.16	0.00	0.00	****	94
21	0	5547	3.20	0.00	0.17	0.00	0.00	****	103
25	0	3712	1.80	0.00	0.21	0.00	0.00	****	123
26	0	2501	1.20	0.00	0.21	0.00	0.00	****	124
27	0	1542	1.50	0.00	0.10	0.00	0.00	****	61
28	0	1534	0.90	0.00	0.17	0.00	0.00	****	102

Ns is number of spontaneous tracks, Nu is number of ^{238}U counts, S is analyzed area of crystal, ρ_s is density of spontaneous tracks, ρ_u is density of ^{238}U counts, σ_T is error for each grain age (1σ), U is uranium density. Uranium concentration for standardization using 91500 standard zircon is 1.583×10^{6} cm^{2}. Epsilon (ε) corresponding to conventional zeta (ζ) value for Fish Canyon Tuff zircon is 35.9±1.2.

Table 1b. U-Pb ages for all analyzed zircon grains obtained from Sp. 151124-1

sample name	Total count ^{206}Pb	^{207}Pb	^{232}Th	^{235}U	^{238}U	Th/U	Isotopic ratios $^{207}Pb/^{206}Pb$	Error 2σ	$^{206}Pb/^{238}U$	Error 2σ	$^{207}Pb/^{235}U$	Error 2σ	Disc.*	Age (Ma) $^{206}Pb/^{238}U$	Error 2σ	$^{207}Pb/^{235}U$	Error 2σ
151124-1 no.1	30	-1	44906	1216	167727	0.47	-0.0387	± #NUM!	0.00012	± 0.000043	-0.0006	± #NUM!	in bad condition	0.76	± 0.28	-0.64	± #NUM!
151124-1 no.2	111	5	145429	2699	372122	0.68	0.0424	± 0.0396	0.00019	± 0.000038	0.0011	± 0.0010	concordant	1.25	± 0.24	1.16	± 1.06
151124-1 no.3	120	58	123218	1629	224546	0.96	0.4702	± 0.1523	0.00035	± 0.000065	0.0227	± 0.0063	-610	2.26	± 0.42	22.8	± 6.37
151124-1 no.4	29	-2	24455	625	86121	0.50	-0.0548	± #NUM!	0.00022	± 0.000082	-0.0016	± #NUM!	in bad condition	1.40	± 0.53	-1.67	± #NUM!
151124-1 no.5	55	3	48866	1158	159626	0.54	0.0568	± 0.0654	0.00023	± 0.000061	0.0018	± 0.0020	concordant	1.46	± 0.40	1.80	± 2.01
151124-1 no.6	50	0	72939	1250	172311	0.74	0.0000	± #NUM!	0.00019	± 0.000054	0.0000	± #NUM!	in bad condition	1.23	± 0.35	0.00	± #NUM!
151124-1 no.7	27	-3	26964	600	82788	0.57	-0.1023	± #NUM!	0.00021	± 0.000082	-0.0030	± #NUM!	in bad condition	1.36	± 0.53	-3.04	± #NUM!
151124-1 no.8	20	-2	18019	460	63451	0.50	-0.0792	± #NUM!	0.00020	± 0.000092	-0.0022	± #NUM!	in bad condition	1.32	± 0.59	-2.26	± #NUM!
151124-1 no.9	46	5	38992	956	131795	0.52	0.0971	± 0.0951	0.00023	± 0.000068	0.0031	± 0.0029	concordant	1.48	± 0.44	3.13	± 2.93
151124-1 no.10	62	18	78352	1249	172172	0.80	0.2894	± 0.1551	0.00023	± 0.000060	0.0093	± 0.0045	-201	1.51	± 0.39	9.44	± 4.52
151124-1 no.11	117	42	70629	1261	173875	0.71	0.3506	± 0.1276	0.00044	± 0.000083	0.0212	± 0.0068	-390	2.83	± 0.53	21.3	± 6.92
151124-1 no.12	45	19	44924	968	133399	0.59	0.4186	± 0.2280	0.00022	± 0.000066	0.0128	± 0.0060	-353	1.43	± 0.43	13.0	± 6.03
151124-1 no.13	61	10	76956	1327	182984	0.74	0.1537	± 0.1069	0.00022	± 0.000056	0.0046	± 0.0030	concordant	1.41	± 0.36	4.69	± 3.06
151124-1 no.14	59	33	27248	821	113208	0.42	0.5550	± 0.2427	0.00034	± 0.000089	0.0259	± 0.0093	-631	2.18	± 0.57	26.0	± 9.45
151124-1 no.15	90	4	86721	1923	265165	0.57	0.0414	± 0.0434	0.00022	± 0.000047	0.0013	± 0.0013	concordant	1.43	± 0.31	1.28	± 1.32
151124-1 no.16	60	10	16839	719	99184	0.30	0.1638	± 0.1122	0.00039	± 0.000103	0.0089	± 0.0057	-1	2.54	± 0.66	9.00	± 5.78
151124-1 no.17	56	10	38042	877	120900	0.55	0.1702	± 0.1181	0.00031	± 0.000082	0.0072	± 0.0046	-3	1.97	± 0.53	7.24	± 4.69
151124-1	20	4	10038	285	39315	0.45	0.1999	± 0.2196	0.00033	± 0.000148	0.0090	± 0.0091	concordant	2.10	± 0.95	9.08	± 9.17

sample name	Total count ^{206}Pb	^{207}Pb	^{232}Th	^{235}U	^{238}U	$\frac{Th}{U}$	Isotopic ratios $\frac{^{207}Pb}{^{206}Pb}$	Error 2σ	$\frac{^{206}Pb}{^{238}U}$	Error 2σ	$\frac{^{207}Pb}{^{235}U}$	Error 2σ	Disc.*	Age (Ma) $\frac{^{206}Pb}{^{238}U}$	Error 2σ	$\frac{^{207}Pb}{^{235}U}$	Error 2σ
no.18																	
151124-1 no.19	57	6	35842	1044	143962	0.44	0.1031	± 0.0887	0.00026	± 0.000069	0.0037	± 0.0030	concordant	1.67	± 0.45	3.73	± 3.07
151124-1 no.20	100	4	33293	702	96815	0.60	0.0411	± 0.0410	0.00068	± 0.000137	0.0038	± 0.0038	concordant	4.36	± 0.88	3.88	± 3.81
151124-1 no.21	144	7	36300	808	111417	0.57	0.0504	± 0.0381	0.00084	± 0.000144	0.0059	± 0.0044	concordant	5.44	± 0.93	5.94	± 4.41
151124-1 no.22	44	7	55409	1189	163980	0.59	0.1655	± 0.1318	0.00017	± 0.000053	0.0040	± 0.0030	concordant	1.13	± 0.34	4.04	± 2.99
151124-1 no.23	70	16	66667	1310	180624	0.65	0.2254	± 0.1247	0.00025	± 0.000061	0.0079	± 0.0040	-117	1.64	± 0.40	8.01	± 4.05
151124-1 no.24	100	3	92967	1614	222501	0.73	0.0328	± 0.0365	0.00029	± 0.000060	0.0013	± 0.0015	concordant	1.89	± 0.38	1.35	± 1.48
151124-1 no.25	57	6	47808	809	111584	0.75	0.0977	± 0.0859	0.00034	± 0.000089	0.0045	± 0.0038	concordant	2.16	± 0.58	4.58	± 3.86
151124-1 no.26	23	2	42651	949	130807	0.57	0.0956	± 0.1325	0.00012	± 0.000049	0.0015	± 0.0020	concordant	0.75	± 0.31	1.57	± 2.08
151124-1 no.27	19	1	14930	433	59693	0.44	0.0299	± 0.0803	0.00020	± 0.000095	0.0008	± 0.0022	concordant	1.32	± 0.61	0.86	± 2.27
151124-1 no.28	36	11	28818	669	92195	0.55	0.2950	± 0.2062	0.00025	± 0.000085	0.0103	± 0.0064	-108	1.63	± 0.55	10.4	± 6.44
151124-1 no.29	36	35	43835	863	118980	0.64	0.9296	± 0.4438	0.00020	± 0.000067	0.0257	± 0.0091	-1149	1.29	± 0.43	25.7	± 9.18
151124-1 no.30	188	7	217620	3679	507208	0.75	0.0354	± 0.0277	0.00024	± 0.000036	0.0012	± 0.0009	concordant	1.56	± 0.23	1.20	± 0.93
													Average ± 2SD	**1.83**	**± 1.94**	**6.45**	**± 16.07**
													Weighted average	**1.50**	**± 0.07**	**2.38**	**± 0.46**

Th/U is ratio of thorium and uranium concentrations. Disc.* is degree of discordance. Positive and negative values mean left- and right-side significant offsets from concordia line, respectively.

Table 1c. U-Pb ages for adopted zircon grains obtained from Sp. 151124-1

sample name	Total count ^{206}Pb	^{207}Pb	^{232}Th	^{235}U	^{238}U	Th/U	Isotopic ratios $^{207}Pb/^{206}Pb$	Error 2σ	$^{206}Pb/^{238}U$	Error 2σ	$^{207}Pb/^{235}U$	Error 2σ	Disc.*	Age (Ma) $^{206}Pb/^{238}U$	Error 2σ	$^{207}Pb/^{235}U$	Error 2σ
151124-1 no.2	111	5	145429	2699	372122	0.68	0.04244	± 0.03962	0.00019±	0.00004	0.00114±	0.00104	concordant	1.25	± 0.24	1.16	± 1.06
151124-1 no.5	55	3	48866	1158	159626	0.54	0.05679	± 0.06536	0.00023±	0.00006	0.00177±	0.00199	concordant	1.46	± 0.40	1.80	± 2.01
151124-1 no.9	46	5	38992	956	131795	0.52	0.09711	± 0.09506	0.00023±	0.00007	0.00308±	0.00289	concordant	1.48	± 0.44	3.13	± 2.93
151124-1 no.13	61	10	76956	1327	182984	0.74	0.15365	± 0.10695	0.00022±	0.00006	0.00463±	0.00302	concordant	1.41	± 0.36	4.69	± 3.06
151124-1 no.15	90	4	86721	1923	265165	0.57	0.04136	± 0.04337	0.00022±	0.00005	0.00127±	0.00130	concordant	1.43	± 0.31	1.28	± 1.32
151124-1 no.18	20	4	10038	285	39315	0.45	0.19991	± 0.21958	0.00033±	0.00015	0.00899±	0.00907	concordant	2.10	± 0.95	9.08	± 9.17
151124-1 no.19	57	6	35842	1044	143962	0.44	0.10311	± 0.08866	0.00026±	0.00007	0.00368±	0.00303	concordant	1.67	± 0.45	3.73	± 3.07
151124-1 no.22	44	7	55409	1189	163980	0.59	0.16549	± 0.13181	0.00017±	0.00005	0.00399±	0.00295	concordant	1.13	± 0.34	4.04	± 2.99
151124-1 no.26	23	2	42651	949	130807	0.57	0.09557	± 0.13253	0.00012±	0.00005	0.00154±	0.00205	concordant	0.75	± 0.31	1.57	± 2.08
151124-1 no.27	19	1	14930	433	59693	0.44	0.02991	± 0.08035	0.00020±	0.00010	0.00085±	0.00224	concordant	1.32	± 0.61	0.86	± 2.27
151124-1 no.30	188	7	217620	3679	507208	0.75	0.03539	± 0.02769	0.00024±	0.00004	0.00118±	0.00091	concordant	1.56	± 0.23	1.20	± 0.93
													Average ± 2SD	**1.42**	**± 0.67**	**2.96**	**± 4.86**
													Weighted average	**1.34**	**± 0.10**	**1.59**	**± 0.52**

Th/U is ratio of thorium and uranium concentrations. Disc.* is degree of discordance. Positive and negative values mean left- and right-side significant offsets from concordia line, respectively.

Table 2a. Fission-track grain ages of zircons obtained from Sp. 151205-1

No.	Ns	Nu	S $\times 10^{-5}$ (cm^2)	ρ_s $\times 10^{-7}$ (cm^{-2})	ρ_u $\times 10^{-9}$ (cm^{-2})	Ns / Nu $\times 10^3$	T (Ma)	σ_T (Ma)	U (ppm)
26	1	4332	1.20	8.33	0.36	0.23	1.27	1.27	223
29	1	4497	2.40	4.17	0.19	0.22	1.22	1.22	116
25	2	11738	4.40	4.55	0.27	0.17	0.94	0.66	165
14	2	13613	2.10	9.52	0.65	0.15	0.81	0.57	400
24	2	14724	2.40	8.33	0.61	0.14	0.75	0.53	379
16	1	7766	1.40	7.14	0.55	0.13	0.71	0.71	342
3	1	8702	1.50	6.67	0.58	0.11	0.63	0.63	358
12	3	26218	5.40	5.56	0.49	0.11	0.63	0.36	300
1	2	18523	3.60	5.56	0.51	0.11	0.59	0.42	318
9	1	12254	4.80	2.08	0.26	0.08	0.45	0.45	158
19	1	12642	4.00	2.50	0.32	0.08	0.43	0.43	195
2	1	13806	3.00	3.33	0.46	0.07	0.40	0.40	284
13	1	15956	4.80	2.08	0.33	0.06	0.34	0.34	205
5	2	33215	2.40	8.33	1.38	0.06	0.33	0.23	854
30	1	17301	3.60	2.78	0.48	0.06	0.32	0.32	297
11	1	19107	2.80	3.57	0.68	0.05	0.29	0.29	421
20	1	19929	3.60	2.78	0.55	0.05	0.28	0.28	342
8	2	40901	7.20	2.78	0.57	0.05	0.27	0.19	351
21	1	26295	5.10	1.96	0.52	0.04	0.21	0.21	318
4	0	15359	4.00	0.00	0.38	0.00	0.00	****	237
6	0	9175	2.40	0.00	0.38	0.00	0.00	****	236
7	0	11898	2.10	0.00	0.57	0.00	0.00	****	350
10	0	30934	6.60	0.00	0.47	0.00	0.00	****	289
15	0	13184	3.40	0.00	0.39	0.00	0.00	****	239
17	0	6396	2.60	0.00	0.25	0.00	0.00	****	152
18	0	13421	3.00	0.00	0.45	0.00	0.00	****	276
22	0	11964	2.40	0.00	0.50	0.00	0.00	****	308
23	0	12066	3.00	0.00	0.40	0.00	0.00	****	248
27	0	9275	1.80	0.00	0.52	0.00	0.00	****	318
28	0	11404	1.80	0.00	0.63	0.00	0.00	****	391

Ns is number of spontaneous tracks, Nu is number of ^{238}U counts, S is analyzed area of crystal, ρ_s is density of spontaneous tracks, ρ_u is density of ^{238}U counts, σ_T is error for each grain age (1σ), U is uranium density. Uranium concentration for standardization using 91500 standard zircon is 1.529×10^6 cm^2. Epsilon (ε) corresponding to conventional zeta (ζ) value for Fish Canyon Tuff zircon is 35.9 ± 1.8.

Table 2b. U-Pb ages for all analyzed zircon grains obtained from Sp. 151205-1

sample name	Total count ^{206}Pb	^{207}Pb	^{232}Th ^{235}U ^{238}U Th/U	Isotopic ratios $^{207}Pb/^{206}Pb$	Error 2σ	$^{206}Pb/^{238}U$	Error 2σ	$^{207}Pb/^{235}U$	Error 2σ	Disc.*	Age (Ma) $^{206}Pb/^{238}U$	Error 2σ	$^{207}Pb/^{235}U$	Error 2σ
151205-1 no.1	22	-3	13039621362945030.77	-0.1467	±#NUM!	0.00005	±0.000021	-0.0010	±#NUM!	in bad condition	0.31	±0.13	-0.98	±#NUM!
151205-1 no.2	64	36	13072120952888070.79	0.5668	±0.2360	0.00014	±0.000037	0.0112	±0.0038	-678	0.92	±0.24	11.30	±3.89
151205-1 no.3	37	4	91158 26173608950.44	0.1012	±0.1090	0.00007	±0.000022	0.0009	±0.0010	concordant	0.43	±0.14	0.95	±0.98
151205-1 no.4	50	3	47383 17322388710.35	0.0634	±0.0731	0.00014	±0.000039	0.0012	±0.0013	concordant	0.88	±0.25	1.21	±1.35
151205-1 no.5	58	11	34490453657397290.81	0.1940	±0.1257	0.00005	±0.000014	0.0014	±0.0008	-42	0.33	±0.09	1.39	±0.83
151205-1 no.6	20	4	61475 18342528680.42	0.1748	±0.1999	0.00005	±0.000023	0.0013	±0.0013	concordant	0.34	±0.15	1.28	±1.36
151205-1 no.7	14	0	17042324583389750.88	0.0000	±0.0000	0.00003	±0.000014	0.0000	±0.0000	366660458	0.17	±0.09	0.00	±0.00
151205-1 no.8	52	16	15432324513379070.80	0.3074	±0.1750	0.00010	±0.000028	0.0043	±0.0021	-202	0.65	±0.18	4.31	±2.18
151205-1 no.9	41	7	42611 10511448740.51	0.1722	±0.1392	0.00019	±0.000059	0.0044	±0.0033	concordant	1.20	±0.38	4.47	±3.36
151205-1 no.1035		4	47155 21312937980.28	0.1030	±0.1142	0.00008	±0.000026	0.0011	±0.0012	concordant	0.49	±0.17	1.10	±1.17
151205-1 no.1116		2	22625932144431580.89	0.1254	±0.1882	0.00002	±0.000012	0.0004	±0.0006	concordant	0.15	±0.08	0.41	±0.58
151205-1 no.1251		9	12515120372809100.78	0.1670	±0.1232	0.00012	±0.000034	0.0027	±0.0019	concordant	0.76	±0.22	2.76	±1.90
151205-1 no.1316		3	39633 13131810520.38	0.2105	±0.2514	0.00006	±0.000029	0.0017	±0.0018	concordant	0.37	±0.19	1.69	±1.84
151205-1 no.1459		6	13660827333767800.63	0.1071	±0.0891	0.00010	±0.000027	0.0015	±0.0012	concordant	0.66	±0.18	1.53	±1.22
151205-1 no.1514		0	83480 17122360360.62	0.0139	±0.0628	0.00004	±0.000021	0.0001	±0.0003	concordant	0.25	±0.13	0.08	±0.34
151205-1 no.1615		7	17129627283761190.79	0.4358	±0.4072	0.00003	±0.000013	0.0016	±0.0012	-53	0.17	±0.09	1.58	±1.24
151205-1 no.179		7	33056 10621464350.39	0.7375	±0.7366	0.00004	±0.000027	0.0042	±0.0032	-214	0.27	±0.18	4.30	±3.28
151205-1 no.1811		0	13219420472821770.82	-0.0174	±#NUM!	0.00003	±0.000016	-0.0001	±#NUM!	in bad condition	0.17	±0.10	-0.06	±#NUM!
151205-1 no.1925		10	75119 14762035750.64	0.4187	±0.3099	0.00008	±0.000032	0.0045	±0.0028	-197	0.51	±0.21	4.60	±2.88
151205-1 no.2017		3	14949722743135830.83	0.1790	±0.2246	0.00003	±0.000017	0.0008	±0.0010	concordant	0.22	±0.11	0.86	±1.00
151205-1 no.2123		0	13089020932886510.79	0.0000	±#NUM!	0.00005	±0.000022	0.0000	±#NUM!	in bad condition	0.34	±0.14	0.00	±#NUM!
151205-1 no.2234		18	11972421532969170.70	0.5216	±0.3057	0.00007	±0.000026	0.0053	±0.0026	-450	0.48	±0.17	5.39	±2.60
151205-1 no.2320		8	94809 18992617970.63	0.4002	±0.3354	0.00005	±0.000022	0.0027	±0.0019	-104	0.32	±0.14	2.75	±1.96
151205-1 no.2461		3	23009036535036240.80	0.0518	±0.0594	0.00008	±0.000021	0.0006	±0.0006	concordant	0.51	±0.13	0.57	±0.64
151205-1 no.2542		31	79010 14151951370.71	0.7275	±0.3446	0.00014	±0.000044	0.0141	±0.0052	-854	0.91	±0.28	14.23	±5.30
151205-1 no.2610		0	62965 15052074520.53	0.0000	±0.0001	0.00003	±0.000020	0.0000	±0.0000	75491473	0.21	±0.13	0.00	±0.00
151205-1 no.2724		-2	14801224133326710.78	-0.0755	±#NUM!	0.00005	±0.000019	-0.0005	±#NUM!	in bad condition	0.30	±0.12	-0.49	±#NUM!

sample name	Total count ^{206}Pb	^{207}Pb	^{232}Th	^{235}U	^{238}U	Th/U	Isotopic ratios $^{207}Pb/^{206}Pb$	Error 2σ	$^{206}Pb/^{238}U$	Error 2σ	$^{207}Pb/^{235}U$	Error 2σ	Disc.*	Age (Ma) $^{206}Pb/^{238}U$	Error 2σ	$^{207}Pb/^{235}U$	Error 2σ
151205-1 no.2835		-1	120330	2748	378834	0.55	-0.0342	±#NUM!	0.00006	±0.000021	-0.0003	±#NUM!	in bad condition	0.39	±0.13	-0.29	±#NUM!
151205-1 no.2927		6	32787	818	112848	0.51	0.2383	±0.2099	0.00015	±0.000060	0.0050	±0.0040	concordant	0.99	±0.39	5.10	±4.07
151205-1 no.3044		24	133163	2022	278771	0.83	0.5398	±0.2731	0.00010	±0.000032	0.0077	±0.0032	-551	0.67	±0.20	7.80	±3.25
													Average ± 2SD	**0.48**	**±0.56**	**2.60**	**±7.00**
													Weighted average	**0.33**	**±0.03**	**0.00**	**±0.00**

Th/U is ratio of thorium and uranium concentrations. Disc.* is degree of discordance. Positive and negative values mean left- and right-side significant offsets from concordia line, respectively.

Table 2c. U-Pb ages for adopted zircon grains obtained from Sp. 151205-1

sample name	Total count ^{206}Pb	^{207}Pb	^{232}Th	^{235}U	^{238}U	Th/U	Isotopic ratios $^{207}Pb/^{206}Pb$	Error 2σ	$^{206}Pb/^{238}U$	Error 2σ	$^{207}Pb/^{235}U$	Error 2σ	Disc.*	Age (Ma) $^{206}Pb/^{238}U$	Error 2σ	$^{207}Pb/^{235}U$	Error 2σ
151205-1 no.3	37	4	91158	2617	360895	0.44	0.10117	± 0.10900	0.00007	± 0.00002	0.00094	± 0.00096	concordant	0.43	± 0.14	0.95	± 0.98
151205-1 no.6	20	4	61475	1834	252868	0.42	0.17478	± 0.19987	0.00005	± 0.00002	0.00126	± 0.00134	concordant	0.34	± 0.15	1.28	± 1.36
151205-1 no.10	35	4	47155	2131	293798	0.28	0.10305	± 0.11416	0.00008	± 0.00003	0.00109	± 0.00115	concordant	0.49	± 0.17	1.10	± 1.17
151205-1 no.11	16	2	226259	3214	443158	0.89	0.12537	± 0.18821	0.00002	± 0.00001	0.00040	± 0.00057	concordant	0.15	± 0.08	0.41	± 0.58
151205-1 no.13	16	3	39633	1313	181052	0.38	0.21046	± 0.25141	0.00006	± 0.00003	0.00167	± 0.00182	concordant	0.37	± 0.19	1.69	± 1.84
151205-1 no.15	14	0	83480	1712	236036	0.62	0.01395	± 0.06282	0.00004	± 0.00002	0.00008	± 0.00034	concordant	0.25	± 0.13	0.08	± 0.34
151205-1 no.20	17	3	149497	2274	313583	0.83	0.17899	± 0.22462	0.00003	± 0.00002	0.00085	± 0.00098	concordant	0.22	± 0.11	0.86	± 1.00
													Average ± 2SD	**0.32**	**± 0.24**	**0.91**	**± 1.08**
													Weighted average	**0.26**	**± 0.05**	**0.38**	**± 0.26**

Th/U is ratio of thorium and uranium concentrations. Disc.* is degree of discordance. Positive and negative values mean left- and right-side significant offsets from concordia line, respectively.

Table 3a. Fission-track grain ages of zircons obtained from Sp. 151206-1

No.	Ns	Nu	S $\times 10^{-5}$ (cm^2)	ρ_s $\times 10^{-7}$ (cm^{-2})	ρ_u $\times 10^{-9}$ (cm^{-2})	Ns / Nu $\times 10^3$	T (Ma)	σ_T (Ma)	U (ppm)
28	10	375	0.90	1.11	0.42	26.67	144.34	46.83	26
29	61	2510	1.40	4.36	1.79	24.30	131.68	18.33	111
30	12	3377	1.20	1.00	2.81	3.55	19.42	5.70	174
18	1	911	1.60	0.06	0.57	1.10	6.01	6.02	35
16	1	970	0.70	0.14	1.39	1.03	5.64	5.65	86
17	4	6632	1.70	0.24	3.90	0.60	3.30	1.66	241
7	1	1666	2.20	0.05	0.76	0.60	3.28	3.29	47
1	1	1888	0.70	0.14	2.70	0.53	2.90	2.90	166
4	1	2459	2.40	0.04	1.02	0.41	2.23	2.23	63
23	1	3022	1.60	0.06	1.89	0.33	1.81	1.81	117
25	1	3063	1.20	0.08	2.55	0.33	1.79	1.79	158
14	2	6506	1.50	0.13	4.34	0.31	1.68	1.19	268
2	0	2287	1.50	0.00	1.52	0.00	0.00	****	94
3	0	2098	3.20	0.00	0.66	0.00	0.00	****	40
5	0	1551	0.80	0.00	1.94	0.00	0.00	****	120
6	0	1891	1.50	0.00	1.26	0.00	0.00	****	78
8	0	1407	0.90	0.00	1.56	0.00	0.00	****	97
9	0	1026	1.80	0.00	0.57	0.00	0.00	****	35
10	0	445	0.60	0.00	0.74	0.00	0.00	****	46
11	0	376	0.80	0.00	0.47	0.00	0.00	****	29
12	0	9900	3.20	0.00	3.09	0.00	0.00	****	191
13	0	1914	0.90	0.00	2.13	0.00	0.00	****	131
15	0	1106	1.20	0.00	0.92	0.00	0.00	****	57
19	0	9037	3.50	0.00	2.58	0.00	0.00	****	159
20	0	3154	0.60	0.00	5.26	0.00	0.00	****	324
21	0	3299	1.00	0.00	3.30	0.00	0.00	****	204
22	0	1107	0.80	0.00	1.38	0.00	0.00	****	85
24	0	2514	1.40	0.00	1.80	0.00	0.00	****	111
26	0	1029	1.00	0.00	1.03	0.00	0.00	****	64
27	0	699	3.00	0.00	0.23	0.00	0.00	****	14

Ns is number of spontaneous tracks, Nu is number of ^{238}U counts, S is analyzed area of crystal, ρ_s is density of spontaneous tracks, ρ_u is density of ^{238}U counts, σ_T is error for each grain age (1σ), U is uranium density. Uranium concentration for standardization using 91500 standard zircon is 1.529×10^6 cm^2. Epsilon (ε) corresponding to conventional zeta (ζ) value for Fish Canyon Tuff zircon is 35.8±1.8.

Table 3b. U-Pb ages for all analyzed zircon grains obtained from Sp. 151206-1

sample name	Total count ^{206}Pb	^{207}Pb	^{232}Th	^{235}U	^{238}U	Th/U	Isotopic ratios $^{207}Pb/^{206}Pb$	Error 2σ	$^{206}Pb/^{238}U$	Error 2σ	$^{207}Pb/^{235}U$	Error 2σ	Disc.*	Age (Ma) $^{206}Pb/^{238}U$	Error 2σ	$^{207}Pb/^{235}U$	Error 2σ
151206-1 no.1	52	-1	104794	1369	188777	0.96	-0.0230±	#NUM!	0.00018±	0.000050	-0.0006±	#NUM!	in bad condition	1.14	± 0.32	-0.57	± #NUM!
151206-1 no.2	27	0	27681	599	82551	0.58	0.0000 ±	#NUM!	0.00021±	0.000082	0.0000 ±	#NUM!	in bad condition	1.37	± 0.53	0.00	± #NUM!
151206-1 no.3	7	-1	13787	261	35989	0.66	-0.1745±	#NUM!	0.00012±	0.000094	-0.0029±	#NUM!	in bad condition	0.78	± 0.60	-2.98	± #NUM!
151206-1 no.4	10	3	20762	447	61588	0.58	0.3164 ±	0.4065	0.00010±	0.000066	0.0046 ±	0.0051	concordant	0.67	± 0.43	4.63	± 5.20
151206-1 no.5	41	1	65173	925	127472	0.88	0.0195 ±	0.0440	0.00021±	0.000065	0.0006 ±	0.0012	concordant	1.32	± 0.42	0.56	± 1.25
151206-1 no.6	61	-1	23873	542	74778	0.55	-0.0193±	#NUM!	0.00053±	0.000138	-0.0014±	#NUM!	in bad condition	3.41	± 0.89	-1.43	± #NUM!
151206-1 no.7	164	67	20129	312	42951	0.81	0.4030 ±	0.1170	0.00247±	0.000404	0.1371 ±	0.0379	-467	15.88	± 2.60	130.44	± 37.76
151206-1 no.8	22	1	59576	697	96114	1.07	0.0366 ±	0.0834	0.00014±	0.000063	0.0007 ±	0.0016	concordant	0.93	± 0.40	0.74	± 1.66
151206-1 no.9	14	0	17617	248	34134	0.89	0.0137 ±	0.0618	0.00027±	0.000144	0.0005 ±	0.0023	concordant	1.75	± 0.93	0.52	± 2.34
151206-1 no.10	57	23	24512	339	46752	0.90	0.3997 ±	0.1968	0.00079±	0.000213	0.0436 ±	0.0189	-350	5.10	± 1.37	43.34	± 19.02
151206-1 no.11	17	-2	15591	232	32042	0.84	-0.1191±	#NUM!	0.00033±	0.000165	-0.0055±	#NUM!	in bad condition	2.15	± 1.06	-5.58	± #NUM!
151206-1 no.12	57	3	57324	1166	160764	0.61	0.0559 ±	0.0643	0.00023±	0.000061	0.0017 ±	0.0020	concordant	1.46	± 0.40	1.77	± 1.99
151206-1 no.13	49	3	48038	969	133667	0.62	0.0608 ±	0.0723	0.00024±	0.000068	0.0020 ±	0.0023	concordant	1.52	± 0.44	2.00	± 2.32
151206-1 no.14	64	5	77213	1693	233444	0.57	0.0840 ±	0.0753	0.00018±	0.000045	0.0020 ±	0.0018	concordant	1.13	± 0.29	2.06	± 1.78
151206-1 no.15	22	-2	27117	386	53163	0.88	-0.0907±	#NUM!	0.00026±	0.000114	-0.0033±	#NUM!	in bad condition	1.70	± 0.73	-3.36	± #NUM!
151206-1 no.16	68	6	32740	634	87351	0.65	0.0925 ±	0.0765	0.00050±	0.000125	0.0064 ±	0.0051	concordant	3.25	± 0.80	6.52	± 5.20
151206-1 no.17	124	6	88798	1901	262106	0.58	0.0447 ±	0.0386	0.00030±	0.000057	0.0019 ±	0.0016	concordant	1.96	± 0.37	1.90	± 1.62
151206-1 no.18	16	1	17012	281	38695	0.76	0.0501 ±	0.1148	0.00026±	0.000133	0.0018 ±	0.0041	concordant	1.70	± 0.86	1.84	± 4.12
151206-1 no.19	60	2	57933	1241	171044	0.58	0.0331 ±	0.0476	0.00023±	0.000059	0.0010 ±	0.0015	concordant	1.45	± 0.38	1.04	± 1.48
151206-1 no.20	111	4	194068	2273	313468	1.07	0.0375 ±	0.0373	0.00023±	0.000045	0.0012 ±	0.0012	concordant	1.47	± 0.29	1.19	± 1.17
151206-1 no.21	66	6	81979	1483	204429	0.69	0.0899 ±	0.0767	0.00021±	0.000052	0.0026 ±	0.0021	concordant	1.34	± 0.34	2.62	± 2.15
151206-1 no.22	17	-2	23133	571	78707	0.51	-0.1035±	#NUM!	0.00014±	0.000068	-0.0020±	#NUM!	in bad condition	0.91	± 0.44	-2.04	± #NUM!
151206-1 no.23	23	-1	60190	795	109593	0.95	-0.0341±	#NUM!	0.00014±	0.000057	-0.0006±	#NUM!	in bad condition	0.88	± 0.37	-0.65	± #NUM!
151206-1 no.24	30	20	52261	748	103086	0.87	0.6636 ±	0.3800	0.00019±	0.000070	0.0174 ±	0.0079	-642	1.22	± 0.45	17.51	± 7.97
151206-1 no.25	39	-3	56247	821	113261	0.86	-0.0853±	#NUM!	0.00022±	0.000072	-0.0026±	#NUM!	in bad condition	1.44	± 0.47	-2.68	± #NUM!
151206-1 no.26	15	1	31358	408	56315	0.96	0.0535 ±	0.1227	0.00017±	0.000088	0.0012 ±	0.0028	concordant	1.09	± 0.57	1.27	± 2.83
151206-1 no.27	7	1	4411	89	12246	0.62	0.0802 ±	0.2153	0.00039±	0.000287	0.0043 ±	0.0112	concordant	2.51	± 1.85	4.36	± 11.27
151206-1 no.28	331	23	4715	151	20859	0.39	0.0699 ±	0.0299	0.01023±	0.001240	0.0986 ±	0.0442	concordant	65.58	± 7.99	95.46	± 43.94
151206-1 no.29	2308	109	22945	858	118303	0.33	0.0468 ±	0.0092	0.01257±	0.000813	0.0811 ±	0.0172	concordant	80.53	± 5.24	79.19	± 17.35
151206-1 no.30	3309	189	47301	1293	178269	0.46	0.0565 ±	0.0085	0.01196±	0.000723	0.0933 ±	0.0156	concordant	76.65	± 4.66	90.54	± 15.75
													Average ± 2SD	**9.41**	**± 44.50**	**15.67**	**± 70.06**
													Weighted average	**1.42**	**± 0.09**	**1.67**	**± 0.50**

Th/U is ratio of thorium and uranium concentrations. Disc.* is degree of discordance. Positive and negative values mean left- and right-side significant offsets from concordia line, respectively.

Table 3c. U-Pb ages for adopted zircon grains obtained from Sp. 151206-1

sample name	Total count					Isotopic ratios						Disc.*	Age (Ma)			
	^{206}Pb	^{207}Pb	^{232}Th	^{235}U	^{238}U	$\frac{Th}{U}$	$\frac{^{207}Pb}{^{206}Pb}$	Error 2σ	$\frac{^{206}Pb}{^{238}U}$	Error 2σ	$\frac{^{207}Pb}{^{235}U}$	Error 2σ		$\frac{^{206}Pb}{^{238}U}$	Error 2σ	$\frac{^{207}Pb}{^{235}U}$	Error 2σ
151206-1 no.4	10	3	20762	447	61588	0.58	0.31644	±0.40651	0.00010	±0.00007	0.00457	±0.00513	concordant	0.67	±0.43	4.63	±5.20
151206-1 no.5	41	1	65173	925	127472	0.88	0.01949	±0.04400	0.00021	±0.00007	0.00055	±0.00123	concordant	1.32	±0.42	0.56	±1.25
151206-1 no.8	22	1	59576	697	96114	1.07	0.03663	±0.08340	0.00014	±0.00006	0.00073	±0.00164	concordant	0.93	±0.40	0.74	±1.66
151206-1 no.9	14	0	17617	248	34134	0.89	0.01373	±0.06185	0.00027	±0.00014	0.00051	±0.00230	concordant	1.75	±0.93	0.52	±2.34
151206-1 no.1257		3	57324	1166	160764	0.61	0.05591	±0.06426	0.00023	±0.00006	0.00175	±0.00196	concordant	1.46	±0.40	1.77	±1.99
151206-1 no.1349		3	48038	969	133667	0.62	0.06079	±0.07233	0.00024	±0.00007	0.00197	±0.00228	concordant	1.52	±0.44	2.00	±2.32
151206-1 no.1464		5	77213	1693	233444	0.57	0.08396	±0.07528	0.00018	±0.00004	0.00203	±0.00176	concordant	1.13	±0.29	2.06	±1.78
151206-1 no.1816		1	17012	281	38695	0.76	0.05007	±0.11476	0.00026	±0.00013	0.00182	±0.00407	concordant	1.70	±0.86	1.84	±4.12
151206-1 no.1960		2	57933	1241	171044	0.58	0.03307	±0.04755	0.00023	±0.00006	0.00103	±0.00146	concordant	1.45	±0.38	1.04	±1.48
151206-1 no.20111		4	194068	2273	313468	1.07	0.03748	±0.03727	0.00023	±0.00004	0.00118	±0.00115	concordant	1.47	±0.29	1.19	±1.17
151206-1 no.2166		6	81979	1483	204429	0.69	0.08990	±0.07668	0.00021	±0.00005	0.00258	±0.00212	concordant	1.34	±0.34	2.62	±2.15
151206-1 no.2615		1	31358	408	56315	0.96	0.05345	±0.12272	0.00017	±0.00009	0.00125	±0.00280	concordant	1.09	±0.57	1.27	±2.83
151206-1 no.277		1	4411	89	12246	0.62	0.08018	±0.21526	0.00039	±0.00029	0.00431	±0.01116	concordant	2.51	±1.85	4.36	±11.27
													Average ± 2SD	**1.41**	**±0.89**	**1.89**	**±2.63**
													Weighted average	**1.28**	**±0.12**	**1.28**	**±0.53**

Th/U is ratio of thorium and uranium concentrations. Disc.* is degree of discordance. Positive and negative values mean left- and right-side significant offsets from concordia line, respectively.

Table 4a. Fission-track grain ages of zircons obtained from Sp. 151219-1

No.	Ns	Nu	S $\times 10^{-5}$ (cm^2)	ρ_s $\times 10^{-7}$ (cm^{-2})	ρ_u $\times 10^{-9}$ (cm^{-2})	Ns / Nu $\times 10^3$	T (Ma)	σ_T (Ma)	U (ppm)
2	25	3502	1.80	1.39	0.19	7.14	37.00	7.52	131
9	17	2822	0.80	2.13	0.35	6.02	31.24	7.67	238
5	64	10692	1.20	5.33	0.89	5.99	31.04	4.02	600
22	32	5763	0.80	4.00	0.72	5.55	28.80	5.19	485
20	21	3936	1.40	1.50	0.28	5.34	27.67	6.12	189
28	142	28628	2.00	7.10	1.43	4.96	25.73	2.32	964
14	89	19311	3.00	2.97	0.64	4.61	23.91	2.66	434
16	56	12224	1.00	5.60	1.22	4.58	23.77	3.28	824
30	50	11090	1.60	3.13	0.69	4.51	23.39	3.40	467
17	41	9283	1.20	3.42	0.77	4.42	22.92	3.66	521
21	21	4914	2.10	1.00	0.23	4.27	22.17	4.90	158
6	19	4465	0.60	3.17	0.74	4.26	22.08	5.13	501
8	22	5567	1.20	1.83	0.46	3.95	20.51	4.43	313
10	68	17230	3.60	1.89	0.48	3.95	20.48	2.58	322
19	69	17724	1.40	4.93	1.27	3.89	20.20	2.52	853
27	40	10279	1.80	2.22	0.57	3.89	20.19	3.27	385
12	84	21621	2.00	4.20	1.08	3.89	20.16	2.30	728
15	38	10569	2.40	1.58	0.44	3.60	18.66	3.09	297
11	23	6482	1.20	1.92	0.54	3.55	18.42	3.89	364
1	26	7457	1.40	1.86	0.53	3.49	18.10	3.60	359
4	32	9609	2.00	1.60	0.48	3.33	17.29	3.11	324
26	71	21839	1.80	3.94	1.21	3.25	16.88	2.08	817
23	23	7369	1.00	2.30	0.74	3.12	16.20	3.42	496
18	93	29890	2.40	3.88	1.25	3.11	16.15	1.76	839
24	12	3919	0.90	1.33	0.44	3.06	15.90	4.62	293
25	19	6299	1.20	1.58	0.52	3.02	15.66	3.63	354
29	38	13129	0.90	4.22	1.46	2.89	15.03	2.49	983
13	52	18277	1.50	3.47	1.22	2.85	14.77	2.11	821
3	38	14246	2.10	1.81	0.68	2.67	13.85	2.29	457
7	31	13675	2.40	1.29	0.57	2.27	11.77	2.15	384

Ns is number of spontaneous tracks, Nu is number of ^{238}U counts, S is analyzed area of crystal, ρ_s is density of spontaneous tracks, ρ_u is density of ^{238}U counts, σ_T is error for each grain age (1σ), U is uranium density. Uranium concentration for standardization using 91500 standard zircon is 1.401×10^6 cm^2. Epsilon (ε) corresponding to conventional zeta (ζ) value for Fish Canyon Tuff zircon is 37.1±1.2.

Table 4b. U-Pb ages for all analyzed zircon grains obtained from Sp. 151219-1

sample name	Total count ^{206}Pb	^{207}Pb	^{232}Th	^{235}U	^{238}U	Th/U	Isotopic ratios $^{207}Pb/^{206}Pb$	Error 2σ	$^{206}Pb/^{238}U$	Error 2σ	$^{207}Pb/^{235}U$	Error 2σ	Disc.*	Age (Ma) $^{206}Pb/^{238}U$	Error 2σ	$^{207}Pb/^{235}U$	Error 2σ
151219-1 no.1	640	32	23130	1211	166922	0.19	0.0499 ±	0.0182	0.00247±	0.000212	0.0170 ±	0.0061	concordant	15.9	±1.37	17.1	±6.21
151219-1 no.2	228	14	19930	457	62948	0.44	0.0589 ±	0.0331	0.00233±	0.000319	0.0189 ±	0.0105	concordant	15.0	±2.05	19.0	±10.57
151219-1 no.3	921	45	32386	1697	233984	0.19	0.0484 ±	0.0150	0.00253±	0.000187	0.0169 ±	0.0052	concordant	16.3	±1.21	17.0	±5.26
151219-1 no.4	556	22	18316	1120	154426	0.17	0.0383 ±	0.0170	0.00232±	0.000211	0.0122 ±	0.0054	concordant	14.9	±1.36	12.4	±5.44
151219-1 no.5	1258	69	45964	2251	310407	0.21	0.0539 ±	0.0138	0.00261±	0.000171	0.0194 ±	0.0049	concordant	16.8	±1.10	19.5	±4.95
151219-1 no.6	947	52	53749	1840	253687	0.29	0.0539 ±	0.0157	0.00240±	0.000176	0.0179 ±	0.0051	concordant	15.5	±1.13	18.0	±5.21
151219-1 no.7	663	36	48176	1311	180807	0.37	0.0532 ±	0.0185	0.00236±	0.000200	0.0173 ±	0.0060	concordant	15.2	±1.29	17.4	±6.03
151219-1 no.8	590	35	22839	1139	156979	0.20	0.0585 ±	0.0207	0.00242±	0.000215	0.0195 ±	0.0068	concordant	15.6	±1.39	19.6	±6.88
151219-1 no.9	5283	258	26474	810	111729	0.33	0.0480 ±	0.0068	0.03045±	0.001322	0.2018 ±	0.0312	concordant	193.4	±8.51	186.6	±31.23
151219-1 no.10	539	16	17800	1018	140406	0.18	0.0289 ±	0.0149	0.00247±	0.000228	0.0098 ±	0.0050	concordant	15.9	±1.47	10.0	±5.09
151219-1 no.11	664	37	49270	1380	190236	0.36	0.0546 ±	0.0188	0.00225±	0.000190	0.0169 ±	0.0057	concordant	14.5	±1.22	17.0	±5.82
151219-1 no.12	1408	76	77539	2694	371477	0.29	0.0532 ±	0.0130	0.00244±	0.000153	0.0179 ±	0.0043	concordant	15.7	±0.99	18.0	±4.35
151219-1 no.13	1662	94	102655	3068	422948	0.34	0.0555 ±	0.0123	0.00253±	0.000150	0.0194 ±	0.0042	concordant	16.3	±0.97	19.5	±4.28
151219-1 no.14	890	48	34148	1778	245115	0.19	0.0527 ±	0.0160	0.00234±	0.000175	0.0170 ±	0.0051	concordant	15.0	±1.13	17.1	±5.16
151219-1 no.15	372	20	14535	783	107954	0.19	0.0530 ±	0.0245	0.00222±	0.000242	0.0162 ±	0.0074	concordant	14.3	±1.56	16.3	±7.49
151219-1 no.16	2093	109	170298	4008	552681	0.43	0.0524 ±	0.0111	0.00245±	0.000141	0.0177 ±	0.0038	concordant	15.7	±0.91	17.8	±3.82
151219-1 no.17	1098	48	57637	1958	269947	0.30	0.0443 ±	0.0135	0.00263±	0.000187	0.0161 ±	0.0049	concordant	16.9	±1.20	16.2	±4.95
151219-1 no.18	1483	73	77702	2841	391749	0.28	0.0500 ±	0.0126	0.00244±	0.000156	0.0168 ±	0.0042	concordant	15.7	±1.01	17.0	±4.30
151219-1 no.19	1278	74	75521	2507	345677	0.31	0.0582 ±	0.0147	0.00239±	0.000161	0.0191 ±	0.0048	concordant	15.4	±1.04	19.3	±4.89
151219-1 no.20	405	16	23386	749	103265	0.32	0.0409 ±	0.0209	0.00253±	0.000269	0.0143 ±	0.0072	concordant	16.3	±1.73	14.4	±7.32
151219-1 no.21	268	10	8439	552	76150	0.16	0.0377 ±	0.0245	0.00227±	0.000290	0.0118 ±	0.0076	concordant	14.6	±1.87	11.9	±7.69
151219-1 no.22	1725	84	48445	1867	257382	0.26	0.0491 ±	0.0116	0.00433±	0.000264	0.0293 ±	0.0070	concordant	27.8	±1.70	29.3	±7.10
151219-1 no.23	891	40	38495	1727	238124	0.23	0.0458 ±	0.0152	0.00242±	0.000186	0.0153 ±	0.0050	concordant	15.6	±1.20	15.4	±5.10
151219-1 no.24	501	19	18067	971	133911	0.19	0.0391 ±	0.0184	0.00242±	0.000234	0.0130 ±	0.0061	concordant	15.6	±1.51	13.1	±6.16
151219-1 no.25	563	28	33721	1094	150900	0.31	0.0505 ±	0.0199	0.00241±	0.000222	0.0168 ±	0.0066	concordant	15.5	±1.43	16.9	±6.65
151219-1 no.26	1455	72	34381	2784	383868	0.13	0.0500 ±	0.0127	0.00245±	0.000158	0.0169 ±	0.0043	concordant	15.8	±1.02	17.0	±4.34
151219-1 no.27	548	21	18731	1054	145362	0.18	0.0379 ±	0.0173	0.00243±	0.000227	0.0127 ±	0.0058	concordant	15.7	±1.47	12.9	±5.85
151219-1 no.28	1481	77	75006	2815	388188	0.27	0.0522 ±	0.0129	0.00246±	0.000158	0.0177 ±	0.0044	concordant	15.9	±1.02	17.9	±4.45
151219-1 no.29	1820	84	31864	3379	465863	0.10	0.0466 ±	0.0110	0.00252±	0.000151	0.0162 ±	0.0039	concordant	16.2	±0.97	16.3	±3.90
151219-1 no.30	650	31	27545	1123	154786	0.25	0.0488 ±	0.0182	0.00271±	0.000236	0.0182 ±	0.0068	concordant	17.4	±1.52	18.4	±6.87
													Average ± 2SD	**22.0**	**±64.90**	**22.6**	**±62.32**
													Weighted average	**16.1**	**±0.23**	**17.1**	**±0.98**

Th/U is ratio of thorium and uranium concentrations. Disc.* is degree of discordance. Positive and negative values mean left- and right-side significant offsets from concordia line, respectively.

Table 4c. U-Pb ages for adopted zircon grains obtained from Sp. 151219-1

sample name	Total count ^{206}Pb	^{207}Pb	^{232}Th	^{235}U	^{238}U	Th/U	Isotopic ratios $^{207}Pb/^{206}Pb$	Error 2σ	$^{206}Pb/^{238}U$	Error 2σ	$^{207}Pb/^{235}U$	Error 2σ	Disc.*	Age (Ma) $^{206}Pb/^{238}U$	Error 2σ	$^{207}Pb/^{235}U$	Error 2σ
151219-1 no.1	640	32	23130	1211	166922	0.19	0.04987	± 0.01823	0.00247±	0.00021	0.01699±	0.00613	concordant	15.9	± 1.37	17.1	± 6.21
151219-1 no.2	228	14	19930	457	62948	0.44	0.05888	± 0.03307	0.00233±	0.00032	0.01891±	0.01047	concordant	15.0	± 2.05	19.0	± 10.57
151219-1 no.3	921	45	32386	1697	233984	0.19	0.04838	± 0.01504	0.00253±	0.00019	0.01691±	0.00519	concordant	16.3	± 1.21	17.0	± 5.26
151219-1 no.4	556	22	18316	1120	154426	0.17	0.03827	± 0.01695	0.00232±	0.00021	0.01224±	0.00537	concordant	14.9	± 1.36	12.4	± 5.44
151219-1 no.5	1258	69	45964	2251	310407	0.21	0.05387	± 0.01375	0.00261±	0.00017	0.01940±	0.00488	concordant	16.8	± 1.10	19.5	± 4.95
151219-1 no.6	947	52	53749	1840	253687	0.29	0.05390	± 0.01574	0.00240±	0.00018	0.01787±	0.00515	concordant	15.5	± 1.13	18.0	± 5.21
151219-1 no.7	663	36	48176	1311	180807	0.37	0.05319	± 0.01855	0.00236±	0.00020	0.01733±	0.00596	concordant	15.2	± 1.29	17.4	± 6.03
151219-1 no.8	590	35	22839	1139	156979	0.20	0.05845	± 0.02066	0.00242±	0.00022	0.01951±	0.00680	concordant	15.6	± 1.39	19.6	± 6.88
151219-1 no.10	539	16	17800	1018	140406	0.18	0.02888	± 0.01485	0.00247±	0.00023	0.00985±	0.00503	concordant	15.9	± 1.47	10.0	± 5.09
151219-1 no.11	664	37	49270	1380	190236	0.36	0.05459	± 0.01880	0.00225±	0.00019	0.01693±	0.00574	concordant	14.5	± 1.22	17.0	± 5.82
151219-1 no.12	1408	76	77539	2694	371477	0.29	0.05319	± 0.01296	0.00244±	0.00015	0.01791±	0.00430	concordant	15.7	± 0.99	18.0	± 4.35
151219-1 no.13	1662	94	102655	3068	422948	0.34	0.05550	± 0.01229	0.00253±	0.00015	0.01937±	0.00422	concordant	16.3	± 0.97	19.5	± 4.28
151219-1 no.14	890	48	34148	1778	245115	0.19	0.05272	± 0.01603	0.00234±	0.00017	0.01700±	0.00509	concordant	15.0	± 1.13	17.1	± 5.16
151219-1 no.15	372	20	14535	783	107954	0.19	0.05298	± 0.02454	0.00222±	0.00024	0.01622±	0.00741	concordant	14.3	± 1.56	16.3	± 7.49
151219-1 no.16	2093	109	170298	4008	552681	0.43	0.05239	± 0.01111	0.00245±	0.00014	0.01767±	0.00377	concordant	15.7	± 0.91	17.8	± 3.82
151219-1 no.17	1098	48	57637	1958	269947	0.30	0.04434	± 0.01351	0.00263±	0.00019	0.01605±	0.00489	concordant	16.9	± 1.20	16.2	± 4.95
151219-1 no.18	1483	73	77702	2841	391749	0.28	0.04997	± 0.01259	0.00244±	0.00016	0.01685±	0.00425	concordant	15.7	± 1.01	17.0	± 4.30
151219-1 no.19	1278	74	75521	2507	345677	0.31	0.05818	± 0.01469	0.00239±	0.00016	0.01914±	0.00483	concordant	15.4	± 1.04	19.3	± 4.89
151219-1 no.20	405	16	23386	749	103265	0.32	0.04091	± 0.02086	0.00253±	0.00027	0.01428±	0.00724	concordant	16.3	± 1.73	14.4	± 7.32
151219-1 no.21	268	10	8439	552	76150	0.16	0.03774	± 0.02449	0.00227±	0.00029	0.01181±	0.00761	concordant	14.6	± 1.87	11.9	± 7.69
151219-1 no.23	891	40	38495	1727	238124	0.23	0.04578	± 0.01517	0.00242±	0.00019	0.01526±	0.00503	concordant	15.6	± 1.20	15.4	± 5.10
151219-1 no.24	501	19	18067	971	133911	0.19	0.03910	± 0.01836	0.00242±	0.00023	0.01303±	0.00608	concordant	15.6	± 1.51	13.1	± 6.16
151219-1 no.25	563	28	33721	1094	150900	0.31	0.05055	± 0.01991	0.00241±	0.00022	0.01680±	0.00657	concordant	15.5	± 1.43	16.9	± 6.65
151219-1 no.26	1455	72	34381	2784	383868	0.13	0.04998	± 0.01270	0.00245±	0.00016	0.01686±	0.00429	concordant	15.8	± 1.02	17.0	± 4.34
151219-1 no.27	548	21	18731	1054	145362	0.18	0.03795	± 0.01730	0.00243±	0.00023	0.01274±	0.00578	concordant	15.7	± 1.47	12.9	± 5.85
151219-1 no.28	1481	77	75006	2815	388188	0.27	0.05225	± 0.01292	0.00246±	0.00016	0.01774±	0.00439	concordant	15.9	± 1.02	17.9	± 4.45
151219-1 no.29	1820	84	31864	3379	465863	0.10	0.04662	± 0.01104	0.00252±	0.00015	0.01621±	0.00385	concordant	16.2	± 0.97	16.3	± 3.90
151219-1 no.30	650	31	27545	1123	154786	0.25	0.04880	± 0.01824	0.00271±	0.00024	0.01824±	0.00679	concordant	17.4	± 1.52	18.4	± 6.87
													Average ± 2SD	**15.7**	**± 1.43**	**16.5**	**± 4.98**
													Weighted average	**15.7**	**± 0.23**	**16.7**	**± 1.00**

Th/U is ratio of thorium and uranium concentrations. Disc.* is degree of discordance. Positive and negative values mean left- and right-side significant offsets from concordia line, respectively.

Table 5a. Fission-track grain ages of zircons obtained from Sp. 160116-1

No.	Ns	Nu	S $\times 10^{-5}$ (cm ^{2})	ρ_s $\times 10^{-7}$ (cm^{-2})	ρ_u $\times 10^{-9}$ (cm^{-2})	Ns / Nu $\times 10^3$	T (Ma)	σ_T (Ma)	U (ppm)
23	57	1290	1.20	4.75	1.08	44.19	246.34	34.40	64
30	11	369	1.20	0.92	0.31	29.81	167.22	51.49	18
25	53	2030	1.10	4.82	1.85	26.11	146.69	21.03	110
29	109	4767	1.50	7.27	3.18	22.87	128.65	13.22	189
20	44	2183	3.60	1.22	0.61	20.16	113.54	17.72	36
27	80	4621	1.60	5.00	2.89	17.31	97.64	11.51	172
24	47	2929	1.20	3.92	2.44	16.05	90.55	13.67	146
21	46	3136	2.00	2.30	1.57	14.67	82.83	12.63	93
26	66	4528	1.60	4.13	2.83	14.58	82.31	10.59	169
28	68	4867	1.60	4.25	3.04	13.97	78.92	10.01	181
22	61	5888	2.40	2.54	2.45	10.36	58.61	7.81	146
16	5	13467	5.00	0.10	2.69	0.37	2.11	0.95	161
7	4	10971	2.40	0.17	4.57	0.36	2.07	1.04	273
14	3	11106	3.00	0.10	3.70	0.27	1.53	0.89	221
18	1	4276	1.50	0.07	2.85	0.23	1.33	1.33	170
1	3	15054	4.50	0.07	3.35	0.20	1.13	0.66	199
12	2	11270	4.00	0.05	2.82	0.18	1.01	0.71	168
3	1	6698	2.00	0.05	3.35	0.15	0.85	0.85	200
2	1	11213	3.50	0.03	3.20	0.09	0.51	0.51	191
19	1	15005	3.60	0.03	4.17	0.07	0.38	0.38	249
11	1	17508	4.00	0.02	4.38	0.06	0.32	0.32	261
4	0	4719	2.00	0.00	2.36	0.00	0.00	****	141
5	0	4014	1.80	0.00	2.23	0.00	0.00	****	133
6	0	4097	1.20	0.00	3.41	0.00	0.00	****	204
8	0	2544	1.20	0.00	2.12	0.00	0.00	****	126
9	0	4707	2.00	0.00	2.35	0.00	0.00	****	140
10	0	2815	1.20	0.00	2.35	0.00	0.00	****	140
13	0	10823	3.60	0.00	3.01	0.00	0.00	****	179
15	0	6668	2.40	0.00	2.78	0.00	0.00	****	166
17	0	2620	2.40	0.00	1.09	0.00	0.00	****	65

Ns is number of spontaneous tracks, Nu is number of ^{238}U counts, S is analyzed area of crystal, ρ_s is density of spontaneous tracks, ρ_u is density of ^{238}U counts, σ_T is error for each grain age (1σ), U is uranium density. Uranium concentration for standardization using 91500 standard zircon is 1.583×10^6 cm^2. Epsilon (ε) corresponding to conventional zeta (ζ) value for Fish Canyon Tuff zircon is 35.9 ± 1.2.

Table 5b. U-Pb ages for all analyzed zircon grains obtained from Sp. 160116-1

sample name	Total count ^{206}Pb	^{207}Pb	^{232}Th	^{235}U	^{238}U	Th/U	Isotopic ratios $^{207}Pb/^{206}Pb$	Error 2σ	$^{206}Pb/^{238}U$	Error 2σ	$^{207}Pb/^{235}U$	Error 2σ	Disc.*	Age (Ma) $^{206}Pb/^{238}U$	Error 2σ	$^{207}Pb/^{235}U$	Error 2σ
160116-1 no.1	85	8	62277	1719	237018	0.46	0.0939±	0.0696	0.00023±	0.000052	0.0030±	0.0021	concordant	1.50	± 0.33	3.05	± 2.17
160116-1 no.2	104	24	77060	1701	234594	0.57	0.2306±	0.1047	0.00029±	0.000058	0.0092±	0.0038	-171	1.86	± 0.37	9.30	± 3.87
160116-1 no.3	148	37	87068	1681	231832	0.66	0.2503±	0.0926	0.00041±	0.000071	0.0143±	0.0048	-240	2.67	± 0.46	14.43	± 4.90
160116-1 no.4	85	10	57467	1384	190835	0.53	0.1127±	0.0769	0.00029±	0.000064	0.0045±	0.0029	concordant	1.86	± 0.41	4.55	± 2.96
160116-1 no.5	63	8	54268	1309	180449	0.53	0.1190±	0.0916	0.00023±	0.000058	0.0038±	0.0027	concordant	1.47	± 0.38	3.81	± 2.78
160116-1 no.6	90	6	78074	1721	237279	0.57	0.0709±	0.0581	0.00025±	0.000053	0.0024±	0.0019	concordant	1.59	± 0.34	2.44	± 1.94
160116-1 no.7	131	62	197944	2300	317131	1.09	0.4715±	0.1469	0.00027±	0.000048	0.0175±	0.0046	-627	1.73	± 0.31	17.61	± 4.70
160116-1 no.8	74	76	60753	1377	189868	0.56	1.0254±	0.3384	0.00025±	0.000060	0.0358±	0.0087	-1524	1.63	± 0.39	35.72	± 8.84
160116-1 no.9	368	90	101561	1289	177678	1.00	0.2442±	0.0585	0.00135±	0.000152	0.0454±	0.0103	-288	8.68	± 0.98	45.07	± 10.41
160116-1 no.10	62	10	64306	1216	167619	0.67	0.1543±	0.1073	0.00024±	0.000062	0.0051±	0.0033	concordant	1.55	± 0.40	5.17	± 3.37
160116-1 no.11	238	75	153683	2344	323156	0.83	0.3135±	0.0842	0.00048±	0.000066	0.0208±	0.0050	-395	3.09	± 0.42	20.86	± 5.11
160116-1 no.12	86	34	58491	1567	216079	0.47	0.3926±	0.1601	0.00026±	0.000057	0.0140±	0.0050	-425	1.67	± 0.37	14.15	± 5.01
160116-1 no.13	73	6	52947	1542	212612	0.44	0.0760±	0.0667	0.00022±	0.000053	0.0023±	0.0020	concordant	1.45	± 0.34	2.38	± 2.02
160116-1 no.14	194	31	95795	2033	280319	0.60	0.1573±	0.0615	0.00045±	0.000068	0.0098±	0.0036	-99	2.91	± 0.44	9.90	± 3.66
160116-1 no.15	68	17	75500	1572	216713	0.61	0.2549±	0.1375	0.00020±	0.000050	0.0072±	0.0035	-158	1.31	± 0.32	7.25	± 3.53
160116-1 no.16	54	7	54785	1457	200954	0.48	0.1351±	0.1060	0.00018±	0.000048	0.0033±	0.0024	concordant	1.14	± 0.31	3.33	± 2.46
160116-1 no.17	58	86	41128	554	76380	0.94	1.4658±	0.5016	0.00050±	0.000132	0.1007±	0.0242	-2151	3.21	± 0.85	97.40	± 24.29
160116-1 no.18	69	63	68694	1542	212571	0.56	0.9178±	0.3227	0.00021±	0.000052	0.0266±	0.0070	-1318	1.35	± 0.33	26.66	± 7.12
160116-1 no.19	145	148	133640	2343	323112	0.72	1.0151±	0.2422	0.00029±	0.000050	0.0408±	0.0074	-1645	1.88	± 0.32	40.61	± 7.50
160116-1 no.20	1138	231	13720	329	45341	0.53	0.2018±	0.0306	0.01635±	0.001208	0.4549±	0.0835	-179	104.53	± 7.78	380.68	± 81.41
160116-1 no.21	1883	93	47526	806	111157	0.75	0.0493±	0.0107	0.01103±	0.000700	0.0750±	0.0171	concordant	70.73	± 4.51	73.39	± 17.21
160116-1 no.22	3009	144	74802	1172	161568	0.81	0.0474±	0.0084	0.01213±	0.000688	0.0793±	0.0149	concordant	77.73	± 4.43	77.47	± 15.05
160116-1 no.23	4187	216	29273	616	84995	0.60	0.0513±	0.0076	0.03208±	0.001717	0.2269±	0.0388	concordant	203.58	± 11.06	207.64	± 38.63
160116-1 no.24	4722	313	130697	1647	227035	1.01	0.0657±	0.0083	0.01355±	0.000707	0.1228±	0.0171	-10	86.73	± 4.56	117.63	± 17.24
160116-1 no.25	3132	200	60102	1076	148291	0.71	0.0635±	0.0097	0.01375±	0.000774	0.1204±	0.0201	-2	88.06	± 4.99	115.46	± 20.23
160116-1 no.26	4249	361	82872	1497	206346	0.70	0.0843±	0.0101	0.01341±	0.000712	0.1559±	0.0209	-41	85.88	± 4.59	147.10	± 21.01
160116-1 no.27	4254	256	119746	1726	237972	0.88	0.0598±	0.0082	0.01164±	0.000618	0.0960±	0.0143	0	74.63	± 3.98	93.04	± 14.40
160116-1 no.28	4025	260	108120	1559	214964	0.88	0.0643±	0.0088	0.01219±	0.000654	0.1081±	0.0161	-7	78.14	± 4.21	104.19	± 16.20
160116-1 no.29	6290	409	95922	1745	240658	0.70	0.0646±	0.0073	0.01702±	0.000853	0.1517±	0.0194	-9	108.81	± 5.50	143.46	± 19.47
160116-1 no.30	422	62	6649	168	23201	0.50	0.1458±	0.0402	0.01186±	0.001272	0.2384±	0.0725	-81	75.99	± 8.19	217.09	± 71.07
													Average ± 2SD	**36.58**	**± 101.95**	**68.03**	**± 171.50**
													Weighted average	**2.05**	**± 0.09**	**8.23**	**± 0.77**

Th/U is ratio of thorium and uranium concentrations. Disc.* is degree of discordance. Positive and negative values mean left- and right-side significant offsets from concordia line, respectively.

Table 5c. U-Pb ages for adopted zircon grains obtained from Sp. 160116-1

sample name	Total count					Th/U	Isotopic ratios						Disc.*	Age (Ma)			
	^{206}Pb	^{207}Pb	^{232}Th	^{235}U	^{238}U		$^{207}Pb/^{206}Pb$	Error 2σ	$^{206}Pb/^{238}U$	Error 2σ	$^{207}Pb/^{235}U$	Error 2σ		$^{206}Pb/^{238}U$	Error 2σ	$^{207}Pb/^{235}U$	Error 2σ
160116-1 no.1	85	8	62277	1719	237018	0.46	0.09391	± 0.06961	0.00023±	0.00005	0.00301±	0.00214	concordant	1.50	± 0.33	3.05	± 2.17
160116-1 no.4	85	10	57467	1384	190835	0.53	0.11269	± 0.07694	0.00029±	0.00006	0.00449±	0.00292	concordant	1.86	± 0.41	4.55	± 2.96
160116-1 no.5	63	8	54268	1309	180449	0.53	0.11904	± 0.09157	0.00023±	0.00006	0.00376±	0.00274	concordant	1.47	± 0.38	3.81	± 2.78
160116-1 no.6	90	6	78074	1721	237279	0.57	0.07093	± 0.05814	0.00025±	0.00005	0.00241±	0.00191	concordant	1.59	± 0.34	2.44	± 1.94
160116-1 no.10	62	10	64306	1216	167619	0.67	0.15426	± 0.10728	0.00024±	0.00006	0.00511±	0.00333	concordant	1.55	± 0.40	5.17	± 3.37
160116-1 no.13	73	6	52947	1542	212612	0.44	0.07597	± 0.06672	0.00022±	0.00005	0.00235±	0.00200	concordant	1.45	± 0.34	2.38	± 2.02
160116-1 no.16	54	7	54785	1457	200954	0.48	0.13509	± 0.10605	0.00018±	0.00005	0.00328±	0.00243	concordant	1.14	± 0.31	3.33	± 2.46
													Average ± 2SD	**1.51**	**± 0.43**	**3.53**	**± 2.10**
													Weighted average	**1.48**	**± 0.13**	**3.20**	**± 0.91**

Th/U is ratio of thorium and uranium concentrations. Disc.* is degree of discordance. Positive and negative values mean left- and right-side significant offsets from concordia line, respectively.

Table 6a. Fission-track grain ages of zircons obtained from Sp. 141205-4726

No.	Ns	Nu	S $\times 10^{-5}$ (cm^2)	ρ_s $\times 10^{-7}$ (cm^{-2})	ρ_u $\times 10^{-9}$ (cm^{-2})	Ns / Nu $\times 10^3$	T (Ma)	σ_T (Ma)	U (ppm)
19	4	1536	1.50	2.67	1.02	2.60	6.87	3.46	137
6	8	4979	3.60	2.22	1.38	1.61	4.24	1.51	184
18	2	1410	1.20	1.67	1.18	1.42	3.74	2.66	157
24	2	2038	1.40	1.43	1.46	0.98	2.59	1.84	194
16	3	3340	2.20	1.36	1.52	0.90	2.37	1.37	202
8	2	2812	1.60	1.25	1.76	0.71	1.88	1.33	234
1	4	5765	4.40	0.91	1.31	0.69	1.83	0.92	175
20	3	4692	2.40	1.25	1.96	0.64	1.69	0.98	261
28	1	1660	1.20	0.83	1.38	0.60	1.59	1.59	184
30	2	3394	1.60	1.25	2.12	0.59	1.56	1.10	283
11	1	1814	1.60	0.63	1.13	0.55	1.46	1.46	151
23	1	1880	1.60	0.63	1.18	0.53	1.40	1.41	157
17	3	5952	3.60	0.83	1.65	0.50	1.33	0.77	220
2	4	8760	5.40	0.74	1.62	0.46	1.21	0.61	216
15	2	4467	2.40	0.83	1.86	0.45	1.18	0.84	248
14	1	2309	1.20	0.83	1.92	0.43	1.14	1.15	257
26	1	2379	1.60	0.63	1.49	0.42	1.11	1.11	198
25	1	2845	1.80	0.56	1.58	0.35	0.93	0.93	211
21	1	2883	1.40	0.71	2.06	0.35	0.92	0.92	275
13	1	2974	2.20	0.45	1.35	0.34	0.89	0.89	180
3	3	9247	5.70	0.53	1.62	0.32	0.86	0.50	216
9	1	3178	1.60	0.63	1.99	0.31	0.83	0.83	265
4	0	3120	2.00	0.00	1.56	0.00	0.00	****	208
5	0	3762	2.70	0.00	1.39	0.00	0.00	****	186
7	0	2629	1.60	0.00	1.64	0.00	0.00	****	219
10	0	2321	1.80	0.00	1.29	0.00	0.00	****	172
12	0	3463	1.80	0.00	1.92	0.00	0.00	****	256
22	0	2641	2.00	0.00	1.32	0.00	0.00	****	176
27	0	1710	1.20	0.00	1.43	0.00	0.00	****	190
29	0	1373	1.20	0.00	1.14	0.00	0.00	****	153

Ns is number of spontaneous tracks, Nu is number of ^{238}U counts, S is analyzed area of crystal, ρ_s is density of spontaneous tracks, ρ_u is density of ^{238}U counts, σ_T is error for each grain age (1σ), U is uranium density. Uranium concentration for standardization using 91500 standard zircon is 0.708×10^6 cm^2. Epsilon (ε) corresponding to conventional zeta (ζ) value for Fish Canyon Tuff zircon is 37.3 ± 1.7.

Table 6b. U-Pb ages for all analyzed zircon grains obtained from Sp. 141205-4726

sample name	Total count ^{206}Pb	^{207}Pb	^{232}Th	^{235}U	^{238}U	$\frac{Th}{U}$	Isotopic ratios $\frac{^{207}Pb}{^{206}Pb}$	Error 2σ	$\frac{^{206}Pb}{^{238}U}$	Error 2σ	$\frac{^{207}Pb}{^{235}U}$	Error 2σ	Disc.*	Age (Ma) $\frac{^{206}Pb}{^{238}U}$	Error 2σ	$\frac{^{207}Pb}{^{235}U}$	Error 2σ
Sp.141205-4726 no.1	232	41	662989	9138	1259992	0.69	0.1767	± 0.0048	0.00027	± 0.000037	0.0066	± 0.0021	-147	1.74	± 0.24	6.64	± 2.10
Sp.141205-4726 no.2	323	44	897989	11314	1559992	0.76	0.1362	± 0.0037	0.00030	± 0.000036	0.0057	± 0.0017	-93	1.96	± 0.23	5.76	± 1.76
Sp.141205-4726 no.3	310	87	826989	11314	1559992	0.70	0.2806	± 0.0075	0.00029	± 0.000035	0.0112	± 0.0025	-360	1.88	± 0.23	11.35	± 2.49
Sp.141205-4726 no.4	277	23	633989	10879	1499992	0.56	0.0830	± 0.0022	0.00027	± 0.000034	0.0031	± 0.0013	concordant	1.74	± 0.22	3.13	± 1.32
Sp.141205-4726 no.5	223	10	668989	9719	1339992	0.66	0.0448	± 0.0012	0.00024	± 0.000034	0.0015	± 0.0010	concordant	1.57	± 0.22	1.53	± 0.97
Sp.141205-4726 no.6	222	15	564989	9646	1329992	0.56	0.0676	± 0.0018	0.00024	± 0.000034	0.0023	± 0.0012	concordant	1.58	± 0.22	2.31	± 1.20
Sp.141205-4726 no.7	247	-5	1489989	11459	1579992	1.24	-0.0202	± -0.0005	0.00023	± 0.000031	-0.0006	± #NUM!	in bad condition	1.48	± 0.20	-0.65	± #NUM!
Sp.141205-4726 no.8	345	53	950989	12257	1689992	0.74	0.1536	± 0.0041	0.00030	± 0.000035	0.0063	± 0.0018	-128	1.93	± 0.22	6.40	± 1.78
Sp.141205-4726 no.9	295	14	1129989	13853	1909992	0.78	0.0475	± 0.0013	0.00023	± 0.000028	0.0015	± 0.0008	concordant	1.46	± 0.18	1.50	± 0.80
Sp.141205-4726 no.10	233	49	553989	8993	1239992	0.59	0.2103	± 0.0057	0.00028	± 0.000038	0.0080	± 0.0023	-209	1.77	± 0.24	8.06	± 2.34
Sp.141205-4726 no.11	156	5	446991	7905	1089994	0.54	0.0288	± 0.0009	0.00021	± 0.000034	0.0008	± 0.0008	concordant	1.35	± 0.22	0.84	± 0.79
Sp.141205-4726 no.12	287	3	1019991	13417	1849994	0.72	0.0087	± 0.0003	0.00023	± 0.000028	0.0003	± 0.0003	241	1.46	± 0.18	0.27	± 0.35
Sp.141205-4726 no.13	219	2	719991	9428	1299994	0.72	0.0068	± 0.0002	0.00025	± 0.000034	0.0002	± 0.0004	320	1.59	± 0.22	0.23	± 0.38
Sp.141205-4726 no.14	276	27	1059991	13417	1849994	0.75	0.0960	± 0.0030	0.00022	± 0.000027	0.0029	± 0.0011	-13	1.41	± 0.18	2.90	± 1.13
Sp.141205-4726 no.15	339	29	1099991	12982	1789994	0.80	0.0841	± 0.0026	0.00028	± 0.000032	0.0032	± 0.0012	-1	1.78	± 0.20	3.23	± 1.22
Sp.141205-4726 no.16	230	-5	784991	10589	1459994	0.70	-0.0196	± -0.0006	0.00023	± 0.000032	-0.0006	± #NUM!	in bad condition	1.48	± 0.20	-0.63	± #NUM!
Sp.141205-4726 no.17	227	23	708991	11532	1589994	0.58	0.0991	± 0.0031	0.00021	± 0.000029	0.0028	± 0.0012	-9	1.35	± 0.19	2.87	± 1.21
Sp.141205-4726 no.18	223	28	435991	8195	1129994	0.50	0.1233	± 0.0039	0.00029	± 0.000040	0.0049	± 0.0019	-49	1.86	± 0.26	4.93	± 1.89
Sp.141205-4726 no.19	171	4	549991	7144	984994	0.73	0.0205	± 0.0006	0.00025	± 0.000040	0.0007	± 0.0008	concordant	1.64	± 0.26	0.72	± 0.77
Sp.141205-4726 no.20	307	27	970991	13635	1879994	0.67	0.0863	± 0.0027	0.00024	± 0.000029	0.0028	± 0.0011	-1	1.54	± 0.18	2.86	± 1.12
Sp.141205-4726 no.21	317	27	1139991	14360	1979994	0.75	0.0836	± 0.0026	0.00023	± 0.000028	0.0027	± 0.0010	concordant	1.51	± 0.18	2.71	± 1.06
Sp.141205-4726 no.22	197	6	529991	9211	1269994	0.55	0.0279	± 0.0009	0.00023	± 0.000033	0.0009	± 0.0007	concordant	1.46	± 0.22	0.88	± 0.75
Sp.141205-4726 no.23	159	-3	542991	8195	1129994	0.63	-0.0157	± -0.0005	0.00021	± 0.000033	-0.0004	± #NUM!	in bad condition	1.33	± 0.22	-0.45	± #NUM!
Sp.141205-4726 no.24	228	11	711991	10154	1399994	0.66	0.0461	± 0.0015	0.00024	± 0.000033	0.0015	± 0.0009	concordant	1.53	± 0.21	1.52	± 0.94
Sp.141205-4726 no.25	230	37	780991	11024	1519994	0.67	0.1587	± 0.0050	0.00022	± 0.000030	0.0048	± 0.0016	-113	1.43	± 0.20	4.86	± 1.62
Sp.141205-4726 no.26	232	25	636991	10371	1429994	0.58	0.1056	± 0.0033	0.00024	± 0.000032	0.0034	± 0.0014	-21	1.53	± 0.21	3.47	± 1.41
Sp.141205-4726 no.27	200	17	520991	9936	1369994	0.50	0.0825	± 0.0026	0.00021	± 0.000031	0.0024	± 0.0012	concordant	1.38	± 0.20	2.44	± 1.21
Sp.141205-4726 no.28	224	-15	459991	9646	1329994	0.45	-0.0647	± -0.0020	0.00025	± 0.000034	-0.0022	± #NUM!	in bad condition	1.59	± 0.22	-2.21	± #NUM!
Sp.141205-4726 no.29	216	48	453991	7978	1099994	0.54	0.2199	± 0.0069	0.00029	± 0.000040	0.0086	± 0.0025	-219	1.85	± 0.26	8.73	± 2.56
Sp.141205-4726 no.30	303	18	1049991	14795	2039994	0.67	0.0578	± 0.0018	0.00022	± 0.000026	0.0017	± 0.0008	concordant	1.40	± 0.17	1.74	± 0.83
													Average ± 2SD	**1.59**	**± 0.37**	**2.93**	**± 6.12**
													Weighted average	**1.56**	**± 0.04**	**1.40**	**± 0.17**

Th/U is ratio of thorium and uranium concentrations. Disc.* is degree of discordance. Positive and negative values mean left- and right-side significant offsets from concordia line, respectively.

Table 6c. U-Pb ages for adopted zircon grains obtained from Sp. 141205-4726

sample name	Total count ^{206}Pb	^{207}Pb	^{232}Th	^{235}U	^{238}U	$\frac{Th}{U}$	Isotopic ratios $\frac{^{207}Pb}{^{206}Pb}$	Error 2σ	$\frac{^{206}Pb}{^{238}U}$	Error 2σ	$\frac{^{207}Pb}{^{235}U}$	Error 2σ	Disc.*	Age (Ma) $\frac{^{206}Pb}{^{238}U}$	Error 2σ	$\frac{^{207}Pb}{^{235}U}$	Error 2σ
Sp.141205-4726 no.4	277	23	633989	10879	1499992	0.56	0.08303	± 0.00223	0.00027±	0.00003	0.00309±	0.00130	concordant	1.74	± 0.22	3.13	± 1.32
Sp.141205-4726 no.5	223	10	668989	9719	1339992	0.66	0.04484	± 0.00121	0.00024±	0.00003	0.00150±	0.00095	concordant	1.57	± 0.22	1.53	± 0.97
Sp.141205-4726 no.6	222	15	564989	9646	1329992	0.56	0.06757	± 0.00182	0.00024±	0.00003	0.00227±	0.00118	concordant	1.58	± 0.22	2.31	± 1.20
Sp.141205-4726 no.9	295	14	1129989	13853	1909992	0.78	0.04746	± 0.00128	0.00023±	0.00003	0.00148±	0.00079	concordant	1.46	± 0.18	1.50	± 0.80
Sp.141205-4726 no.11	156	5	446991	7905	1089994	0.54	0.02885	± 0.00091	0.00021±	0.00003	0.00083±	0.00078	concordant	1.35	± 0.22	0.84	± 0.79
Sp.141205-4726 no.19	171	4	549991	7144	984994	0.73	0.02047	± 0.00064	0.00025±	0.00004	0.00071±	0.00076	concordant	1.64	± 0.26	0.72	± 0.77
Sp.141205-4726 no.21	317	27	1139991	14360	1979994	0.75	0.08360	± 0.00263	0.00023±	0.00003	0.00267±	0.00104	concordant	1.51	± 0.18	2.71	± 1.06
Sp.141205-4726 no.22	197	6	529991	9211	1269994	0.55	0.02792	± 0.00088	0.00023±	0.00003	0.00087±	0.00074	concordant	1.46	± 0.22	0.88	± 0.75
Sp.141205-4726 no.24	228	11	711991	10154	1399994	0.66	0.04605	± 0.00145	0.00024±	0.00003	0.00150±	0.00093	concordant	1.53	± 0.21	1.52	± 0.94
Sp.141205-4726 no.27	200	17	520991	9936	1369994	0.50	0.08250	± 0.00260	0.00021±	0.00003	0.00241±	0.00119	concordant	1.38	± 0.20	2.44	± 1.21
Sp.141205-4726 no.30	303	18	1049991	14795	2039994	0.67	0.05776	± 0.00182	0.00022±	0.00003	0.00171±	0.00082	concordant	1.40	± 0.17	1.74	± 0.83
													Average ± 2SD	**1.51**	**± 0.24**	**1.76**	**± 1.61**
													Weighted average	**1.50**	**± 0.06**	**1.51**	**± 0.28**

Th/U is ratio of thorium and uranium concentrations. Disc.* is degree of discordance. Positive and negative values mean left- and right-side significant offsets from concordia line, respectively.

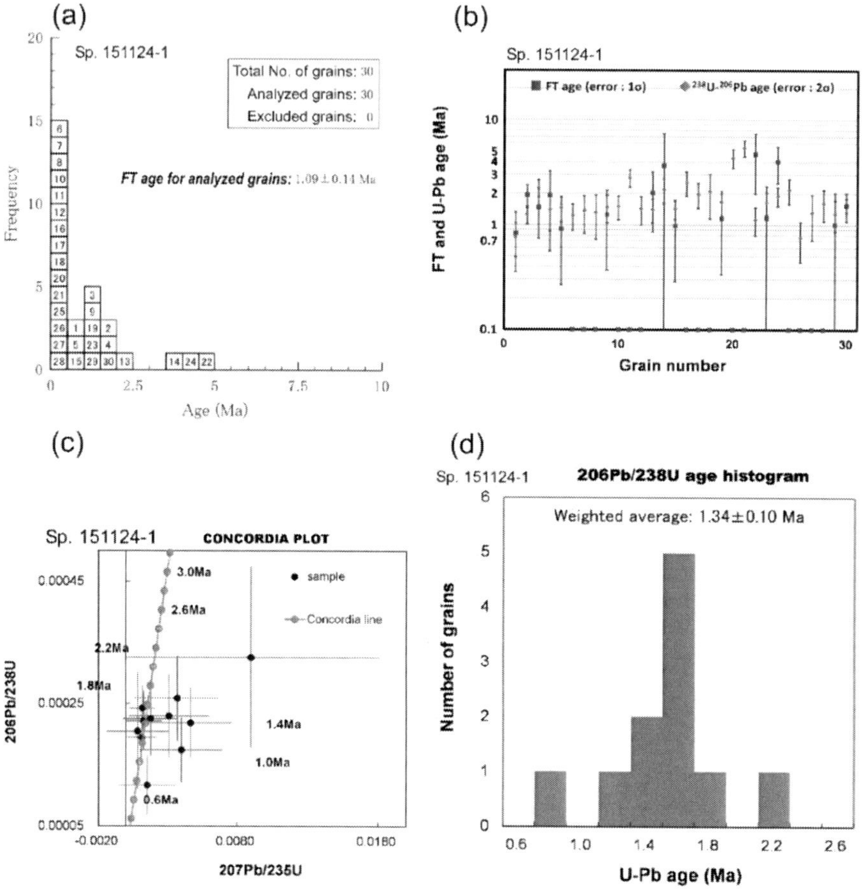

Figure 17. (a) Result of fission-track (FT) dating of zircons obtained from the Shobudani Group. (b) FT and U-Pb age plots for zircons obtained from the Shobudani Group. (c) Concordia plot for U-Pb ages of zircons obtained from the Shobudani Group. (d) Result of U-Pb dating of zircons obtained from the Shobudani Group. See Figure 16 for location (Sp. 151124-1).

Figures 17b–22b compare the FT and U–Pb ages of each of the zircon samples obtained from the Shobudani Group. The upper temperature limit for U–Pb dating is approximately 900 or 700°C under laboratory or natural conditions, respectively, whereas that for zircon FT dating is 240–250°C over a period of 1 million years. Considering this difference, U–Pb ages tend to be larger and have smaller scatter. However, FT dating yielded larger ages for Sp. 151205-1 and 151219-1. It is plausible that the FT age of Sp. 151205-1 is related to the nonequilibrium state of the initial Th–U ratios in the magma chamber and zircon crystals, for which correction is necessary when young samples are being dated. The FT age of Sp. 151219-1 may have been caused by the strong zonation of uranium density in measured crystals.

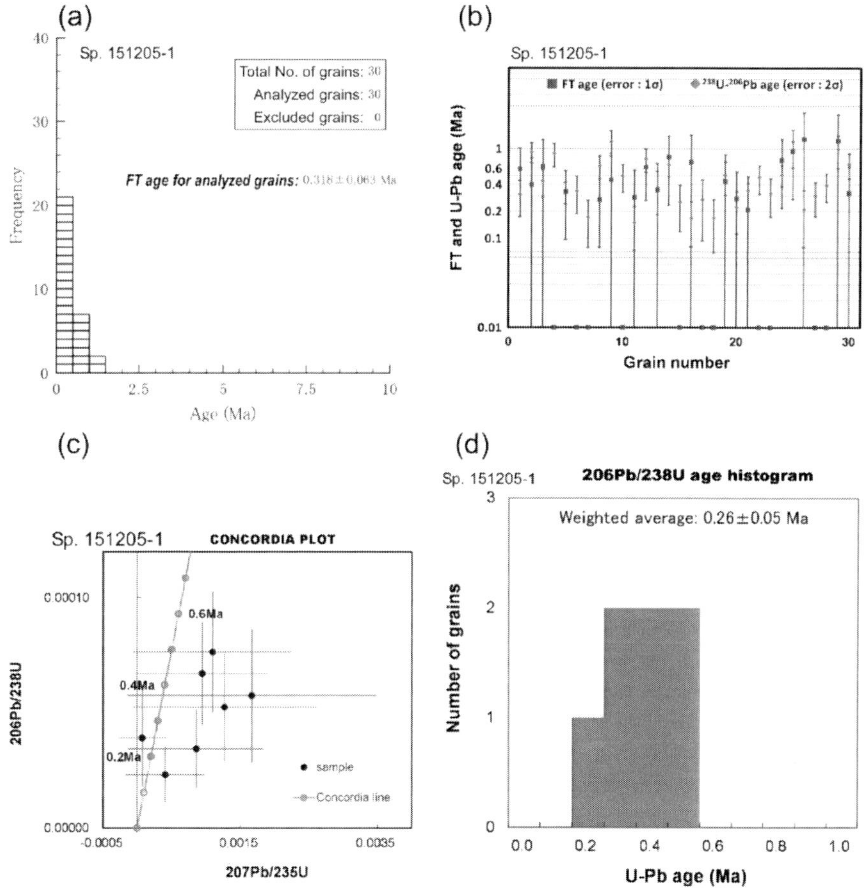

Figure 18. (a) Result of fission-track (FT) dating of zircons obtained from the Shobudani Group. (b) FT and U-Pb age plots for zircons obtained from the Shobudani Group. (c) Concordia plot for U-Pb ages of zircons obtained from the Shobudani Group. (d) Result of U-Pb dating of zircons obtained from the Shobudani Group. See Figure 16 for location (Sp. 151205-1).

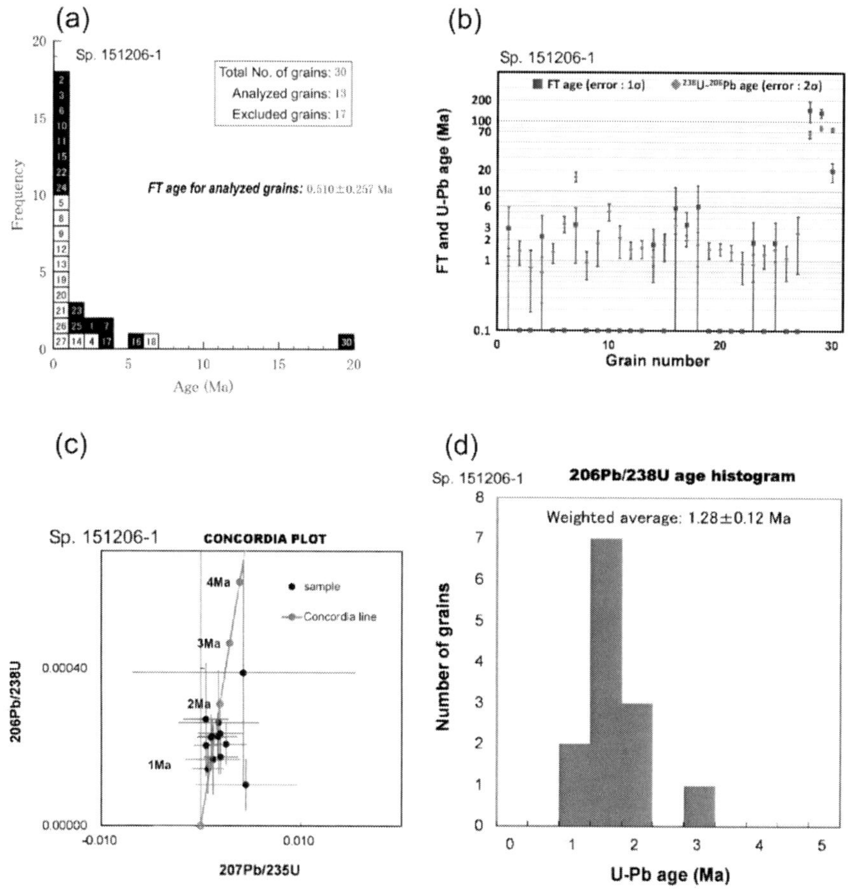

Figure 19. (a) Result of fission-track (FT) dating of zircons obtained from the Shobudani Group. (b) FT and U-Pb age plots for zircons obtained from the Shobudani Group. (c) Concordia plot for U-Pb ages of zircons obtained from the Shobudani Group. (d) Result of U-Pb dating of zircons obtained from the Shobudani Group. See Figure 16 for location (Sp. 151206-1).

The results of U–Pb dating for zircon grains are summarized in Tables 1b–6b (all analyzed grains) and 1c–6c (adopted concordant grains). Based on the concordia plots of the U–Pb zircon ages in Figures 17c–22c, zircon grains with discordant data were excluded, and concordant grains were used to calculate the average. The final results are presented in Figures 17d–22d. The four primary site means (Sp. 141205-4726, 151124-1, 160116-1, and 151206-1) ranging from 1.28 ± 0.12 to 1.50 ± 0.06 Ma are the first numerical ages ever obtained for the Kaseda Formation in the study area. One U–Pb age (0.26 ±

0.05 Ma; Sp. 151205-1) obtained from the Toge Formation is slightly younger than the FT age of Kokawadera ash (0.47 ± 0.08 Ma; Makimoto et al., 2004). Generally speaking, applying the above-mentioned nonequilibrium correction may yield an average U–Pb age of 0.34 Ma.

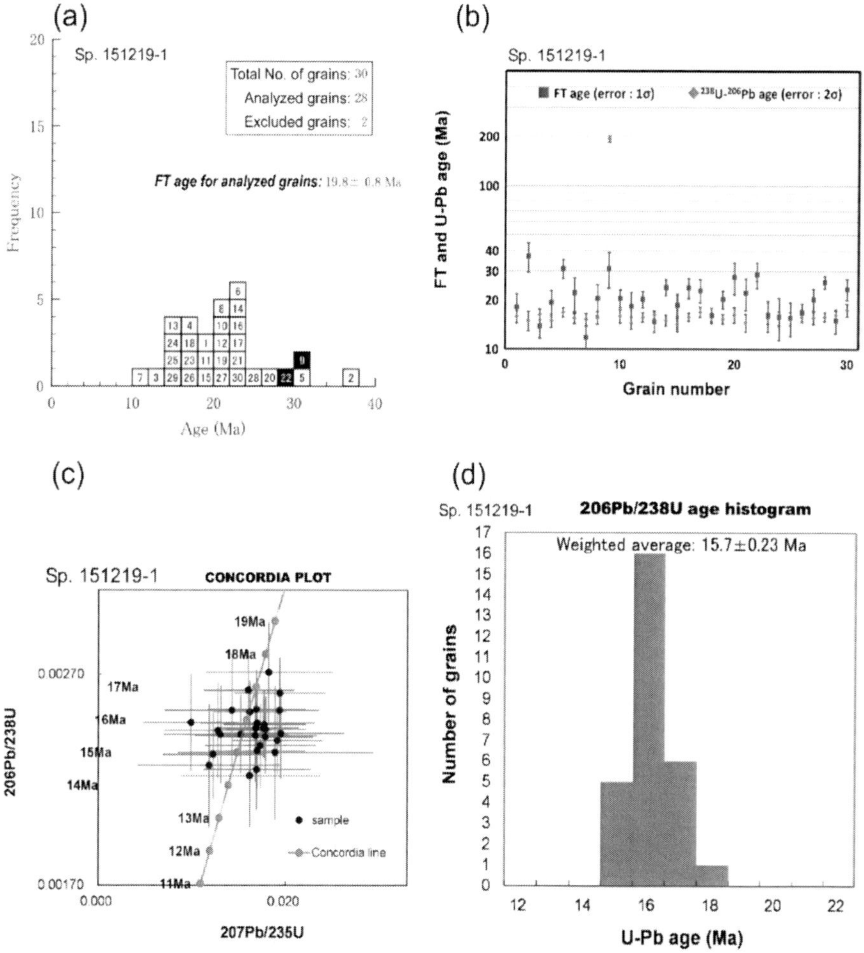

Figure 20. (a) Result of fission-track (FT) dating of zircons obtained from the Shobudani Group. (b) FT and U-Pb age plots for zircons obtained from the Shobudani Group. (c) Concordia plot for U-Pb ages of zircons obtained from the Shobudani Group. (d) Result of U-Pb dating of zircons obtained from the Shobudani Group. See Figure 16 for location (Sp. 151219-1).

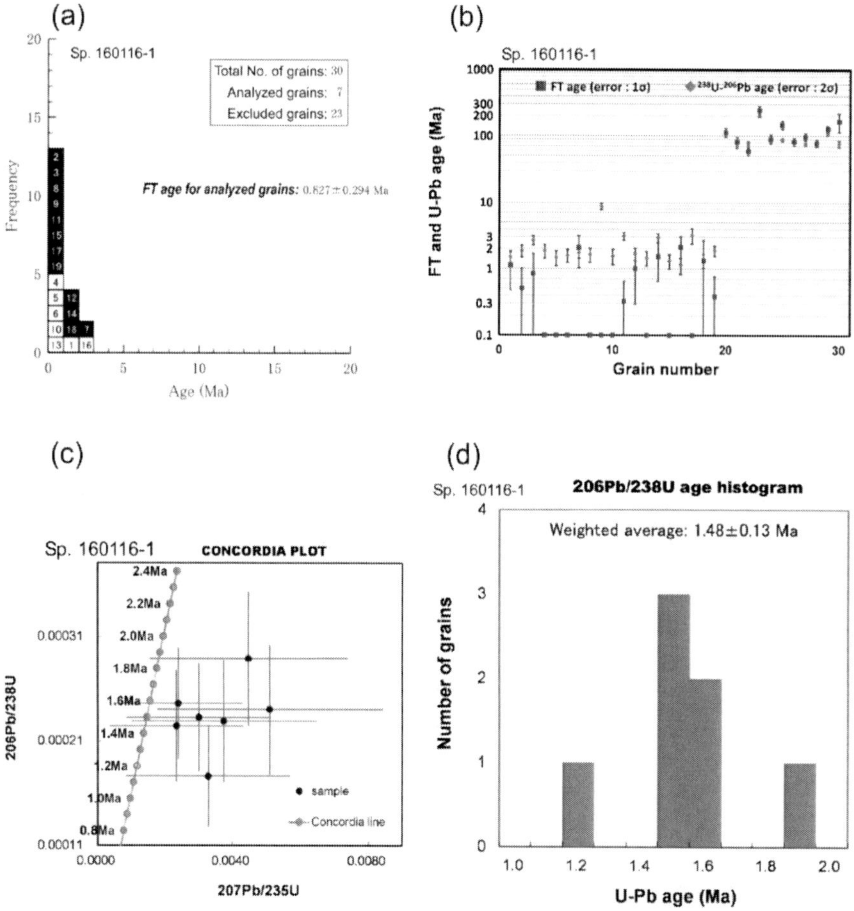

Figure 21. (a) Result of fission-track (FT) dating of zircons obtained from the Shobudani Group. (b) FT and U-Pb age plots for zircons obtained from the Shobudani Group. (c) Concordia plot for U-Pb ages of zircons obtained from the Shobudani Group. (d) Result of U-Pb dating of zircons obtained from the Shobudani Group. See Figure 16 for location (Sp. 160116-1).

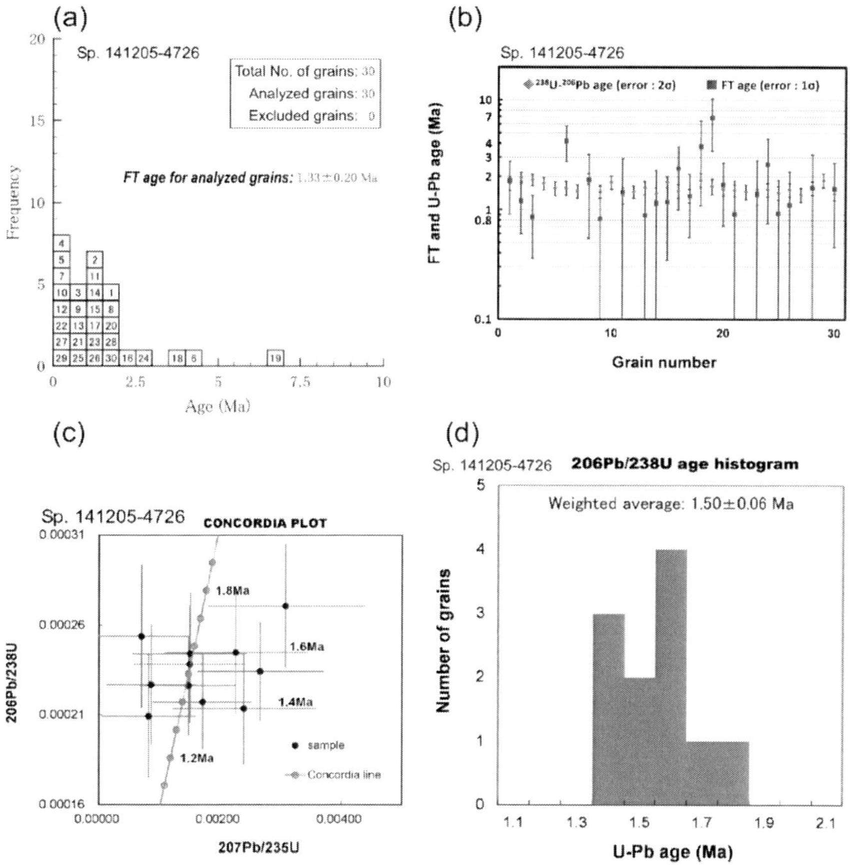

Figure 22. (a) Result of fission-track (FT) dating of zircons obtained from the Shobudani Group. (b) FT and U-Pb age plots for zircons obtained from the Shobudani Group. (c) Concordia plot for U-Pb ages of zircons obtained from the Shobudani Group. (d) Result of U-Pb dating of zircons obtained from the Shobudani Group. See Figure 16 for location (Sp. 141205-4726).

4. Structural Evolution and Temporal Transition of Tectonic Stress

Subsurface structural trends in and around the Osaka sedimentary basin have been determined by interpreting reflection seismic data, and significant fluvial sediments suggestive of paleoenvironmental episodes and causative tectonic events have been dated. To understand the spatiotemporal changes in neotectonic regimes, stratigraphic data of deep boreholes are considered in this section.

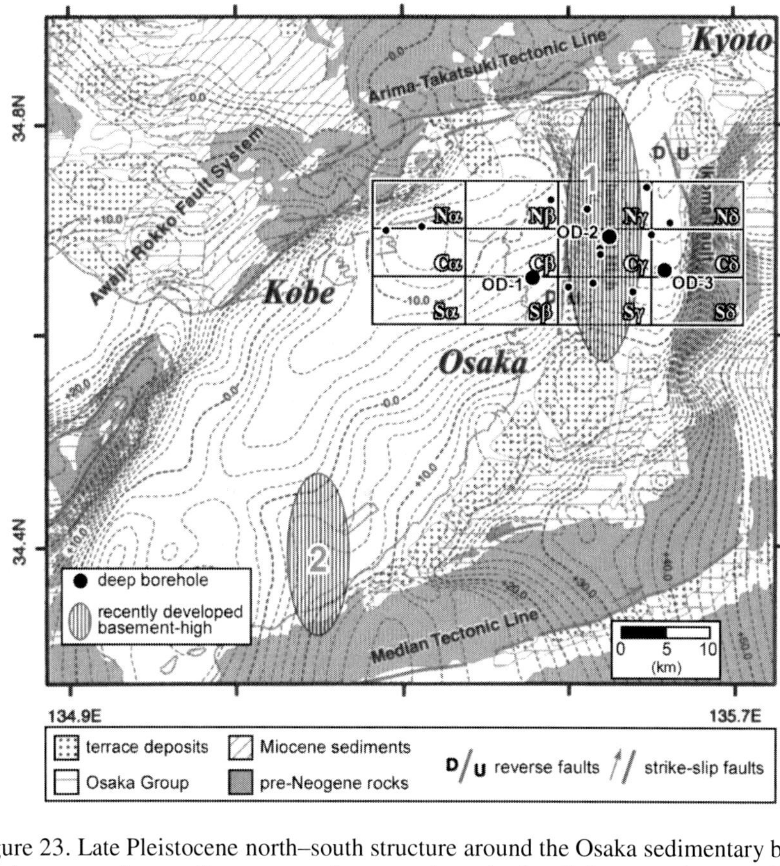

Figure 23. Late Pleistocene north–south structure around the Osaka sedimentary basin. The base map shows a Bouguer gravity anomaly at the 2-mGal contour interval. The Bouguer density is 2670 kg/m^3. A grid delineates the domains for the sediment thickness calculation by Itoh et al. (2000).

Figure 23 shows an index of the area analyzed by Itoh et al. (2000) with a focus on the Plio-Pleistocene subsidence of the Kobe and northern Osaka basins. As they suggested, the northern sector of the Osaka basin consists of eastern and western depressions divided by a north–south basement high (hatched oval 1 in Figure 23), the western side of which is bordered by the Uemachi Fault. Figure 24a and b shows a sediment thickness diagram from the late Pliocene to the early Pleistocene and interval subsidence rates throughout the late Quaternary in the Osaka sedimentary basin after Itoh et al. (2000), respectively. Deep boreholes OD-1, 2, and 3 were drilled on the western sag, basement high, and eastern sag, respectively, and delineate slowing subsidence linked to rising compressive stress since ca. 1 Ma and episodic block motion occurring ca. 0.6–0.5 Ma. In the southern sector of the Osaka basin, Itoh et al. (2001) described a Quaternary tectonic warping (hatched oval 2 in Figure 23) based on interpretations of reflection seismic data and borehole geology. They established a synchronism between the buildup of basement bulges in the northern and southern sectors.

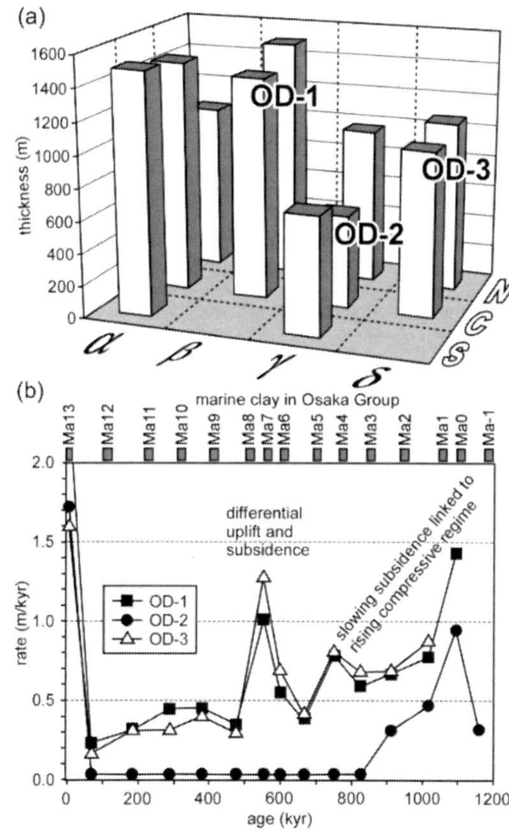

Figure 24. (a) Sediment thickness diagram from late Pliocene to early Pleistocene (from top of basement to Ma 2 marine clay intercalated in Osaka Group) after Itoh et al. (2000). See Figure 23 for the plan view and nomenclature of the analyzed domains. (b) Interval subsidence rates through late Quaternary in Osaka sedimentary basin (Itoh et al., 2000). See Figures 1 and 23 for the locations of the selected boreholes.

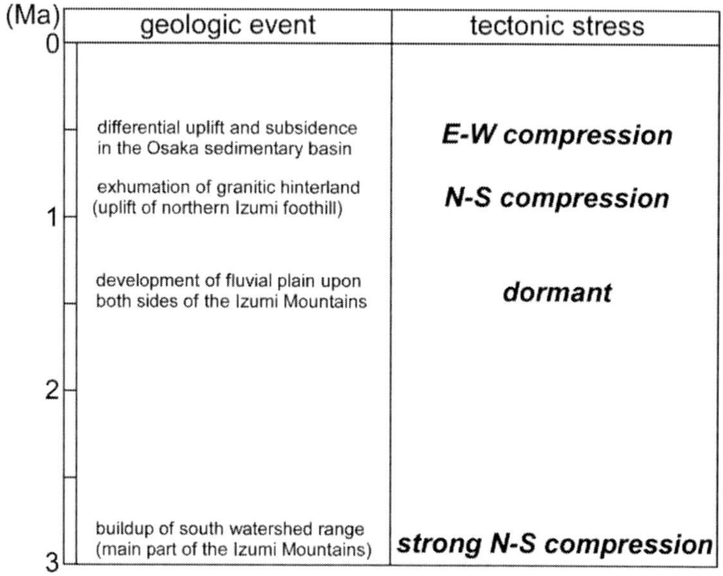

Figure 25. Timeline of tectonic movements around Osaka sedimentary basin since late Pliocene.

Figure 25 summarizes tectonic movements around the Osaka sedimentary basin since the late Pliocene. The most important suggestion for the formation process of the watershed range (Izumi Mountains) was made by Oka (1978) based on a provenance study of gravels exposed on the hillsides of the southern sector of the basin. He related the disappearance of metamorphic clasts derived from the Sanbagawa belt (older) and the upward progressive dominance of volcanic and plutonic clasts (younger) to successive uplift events in the main part and northern foothill of the Izumi range, respectively.

The uplift of the main Izumi range was initially brought about under strong north–south compression, and a survey by Oka (1978) revealed that the event occurred near the horizon of the Mizuma II volcanic ash (ca. 2.9 Ma; Satoguchi and Nagahashi, 2012). As shown in Figure 25, a dormant period has been postulated based on the radiometric ages of the Shobudani Group (Section 3) buried in a fluvial basin along the MTL. Then, the uplift of the northern Izumi flank occurred under north–south compression, which was assigned to Ma 3 marine clay (ca. 0.87 Ma) by Oka (1978). Exhumed Cretaceous granitic rocks have been observed as fault-bounded blocks (Figure 1; Geological Survey of Japan, 2012) and clearly delineated as an east–west structural trend through the gravity analysis discussed in Chapter 2 (Figure 9). The most recent obvious event was east–west compression, under which the compartmentalization of the Osaka sedimentary basin occurred alongside the emergence of basement bulges.

CONCLUSION

This study clarified structural divisions in the Osaka sedimentary basin based on the interpretation of seismic profiles. Its northern sector is characterized by fault-bounded horst and graben structures under prevailing east–west compressive stress, which diminishes southward as active geomorphological features become obscured. The southern sector has been influenced by the activity of the arc-bisecting MTL. According to the fluctuation of the oceanic plates' subduction modes, the MTL has been activated repeatedly with dextral and/or reverse faulting, inevitably causing changes in the stress state in and around the Osaka basin. Incorporating the stratigraphic information of deep boreholes and new radiometric age data for volcanic ashes intercalated in a fluviatile deposit on the MTL, a timeline of tectonic regimes and related geologic events in the Osaka sedimentary basin was obtained in this chapter.

ACKNOWLEDGMENTS

The author thanks the Geological Survey of Japan, AIST (National Institute of Advanced Industrial Science and Technology), and Crisis Management Office, Osaka Prefectural Government for permission to publish this work. The author extends hearty appreciation to Tomotaka Iwata for his continuous support during the course of the comprehensive research and study on the Uemachi Fault Zone (FY2010-2012) organized by the Research and Development Bureau, MEXT, and DPRI, Kyoto University.

REFERENCES

Danhara, T., & Iwano, H. (2009). Determination of zeta values for fission-track age calibration using thermal neutron irradiation at the JRR-3 reactor of JAEA, Japan. *Journal of the Geological Society of Japan,* 115, 141-145.

Disaster Prevention Research Institute (2013). Integrated Research Project for the Uemachi Active Fault System by METI. Uji: DPRI (Kyoto University).

Geological Survey of Japan (1997). Forgoing survey of main active fault in the Kinki Triangle No.18, Ikoma fault system, seismic reflection survey. In K. Shimokawa (Ed.), Open-File Report, No.285. Tsukuba: Geological Survey of Japan, AIST.

Geological Survey of Japan, AIST (Ed.) (2012). Seamless Digital Geological Map of Japan 1:200,000 (July 3, 2012 Version), Research Information Database DB084. Geological Survey of Japan, AIST (National Institute of Advanced Industrial Science and Technology), Tsukuba.

Hirota, Y. (1991). Geology of the Sambagawa metamorphic belt in western Kii Peninsula, Japan. Memoirs of Faculty of Science, Shimane University, 25, 131-142.

Horike, M., Takeuchi, Y., Toriumi, I., Fuzita, T., Yokota, H., & Noda, T. (1995). Seismic reflection survey of the boundary region between the Ikoma Mountains and the Osaka basin. Zisin (Bulletin of Seismological Society of Japan), 48, 37-49.

Horike, M., Takeuchi, Y., Imai, S., Fuzita, T., Yokota, H., Noda, T., & Ikawa, T. (1996). Survey of the subsurface structure in the east of the Osaka basin. Zisin (Bulletin of Seismological Society of Japan), 49, 193-203.

Huzita, K. (1980). Role of the Median Tectonic Line in the Quaternary tectonics of the Japanese Islands. Memoirs of Geological Society of Japan, 18, 129-153.

Huzita, K., & Kasama, T. (1982). Geology of the Osaka-Seihokubu District, with Geological Sheet Map at 1:50,000. Tsukuba: Geological Survey of Japan, 112pp.

Ikebe, N., Iwatsu, J., & Takenaka, J. (1970). Quaternary geology of Osaka with special reference to land subsidence. *Journal of Geoscience,* Osaka City University, 13, 39-98.

Itihara, M., Ichikawa, K., & Yamada, N. (1986). Geology of the Kishiwada District, with Geological Sheet Map at 1:50,000. Tsukuba: Geological Survey of Japan, 148pp.

Itoh, Y., Kusumoto, S., & Takemura, K. (2013a). Characteristic basin formation at terminations of a large transcurrent fault - basin configuration based on gravity and geomagnetic data. In Y. Itoh (Ed.), Mechanism of Sedimentary Basin Formation - Multidisciplinary Approach on Active Plate Margins. Rijeka (Croatia): InTech. http://dx.doi.org/10.5772/56702.

Itoh, Y., Kusumoto, S., & Takemura, K. (2014). Evolutionary process of Beppu Bay in central Kyushu, Japan: a quantitative study of the basin-forming process controlled by plate convergence modes. Earth, Planets and Space, 66, 74. doi:10.1186/1880-5981-66-74.

Itoh, Y., Kusumoto, S., & Takemura, K. (2015). Tectonically controlled asymmetric basin formation and evolution: an example from an active plate margin. In B. Veress, & J. Szigethy (Eds.), Horizons in Earth Science Research, Vol.14. New York: Nova Science Publishers, Inc., pp. 123-141. https://www.novapublishers.com/catalog/product_info.php?products_id=57536.

Itoh, Y., Takemura, K., & Kusumoto, S. (2013b). Neotectonic intra-arc basins within southwest Japan - conspicuous basin-forming process related to differential motion of crustal blocks. In Y. Itoh (Ed.), Mechanism of Sedimentary Basin Formation - Multidisciplinary Approach on Active Plate Margins. Rijeka (Croatia): InTech. http://dx.doi.org/10.5772/56588.

Itoh, Y., Takemura, K., Ishiyama, T., Tanaka, Y., & Iwaki, H. (2000). Basin formation at a contractional bend of a large transcurrent fault: Plio-Pleistocene subsidence of the Kobe and northern Osaka Basins, Japan. Tectonophysics, 321, 327-341.

Itoh, Y., Takemura, K., Kawabata, D., Tanaka, Y., & Nakaseko, K. (2001). Quaternary tectonic warping and strata formation in the southern Osaka Basin inferred from reflection seismic interpretation and borehole sequences. *Journal of Asian Earth Sciences,* 20, 45-58.

Kasahara, K., Aoi, S., Irikura, K., Iwata, T., Kudo, K., Horike, M., Ikawa, T., Kawanaka, T., & Mizohata, S. (1997). Seismic reflection and VSP servey on and around the deep well at Maishima and Kanku-Maejima. Programme and abstracts, the Seismological Society of Japan 1997, No. 2, P36.

Kondo, H., Sugito, N., Yoshioka, T., Tsutsumi, H., & Kimura, H. (2015). Revisited spatial distribution of the Uemachi fault zone, Osaka, revealed by digital elevation models. *Active Fault Research,* 42, 1-34.

Makimoto, H., Miyata, T., Mizuno, K., & Sangawa, A. (2004). Geology of the Kokawa District, with Geological Sheet Map at 1:50,000. Tsukuba: Geological Survey of Japan, AIST, 89pp.

Miyachi, Y., Tainosho, Y., Yoshikawa, T., & Sangawa, A. (1998). Geology of the Osaka-Tonambu District, with Geological Sheet Map at 1:50,000. Tsukuba: Geological Survey of Japan, 113pp.

Oka, Y. (1961). The geomorphology and the crustal movements in the southeastern part of the Osaka Plain. *Geographical Review of Japan*, 34, 523-535.

Oka, Y. (1978). The formation of the Izumi range and the Osaka Group. *The Quaternary Research,* 16, 201-210.

Okada, A., & Togo, M. (2000). Active Faults in Kinki. Tokyo: University of Tokyo Press, 395pp.

Osaka City (1995). Research Report for Uemachi Fault in 1995. http://www.hp1039.jishin.go.jp/danso/OsakaCtyfrm.htm (accessed 21 April 2016).

Osaka Prefecture (2004). Subsurface Structural Survey of the Osaka Plain in the Heisei 15 Fiscal Year. http://www.hp1039.jishin.go.jp/kozo/osaka8 frm.htm (accessed 25 March 2016).

Osaka Prefecture (2005). Subsurface Structural Survey of the Osaka Plain in the Heisei 16 Fiscal Year. http://www.hp1039.jishin.go.jp/kozo/Osaka9 frm.htm (accessed 25 March 2016).

Research and Development Bureau, MEXT, & DPRI, Kyoto University (2012). Reports of Comprehensive Research and Study on the Uemachi Fault Zone (FY2010-2012). http://www.jishin.go.jp/main/chousakenkyuu/ uemachi_juten/(accessed 21 April 2016).

Research Committee on Ground in Kansai (2007). Shin-Kansai Jiban (Ground of Kansai) - From Osaka Plain to Bay Area Rev. ed. Osaka: Association of Research on Geotechnical Information in Kansai, 354pp.

Research Group for Active Faults (1991). The Active Faults in Japan: Sheet Maps and Inventories Rev. ed. Tokyo: University of Tokyo Press, 437pp.

Satoguchi, Y., & Nagahashi, Y. (2012). Tephrostratigraphy of the Pliocene to Middle Pleistocene Series in Honshu and Kyushu Islands, Japan. Island Arc, 21, 149-169.

Wang, C. L., & Maekawa, H. (1997). Albite-biotite zone of the Sanbagawa metamorphic belt in the northwestern part of the Kii Peninsula, Japan. *Journal of Mineralogy, Petrology and Economic Geology*, 92, 43-54.

Chapter 4

TECTONIC IMPLICATION OF BASIN FORMATION ON THE BASIS OF PALEOGEOGRAPHY USING STRATIGRAPHY AND SHALLOW SUBSURFACE STRUCTURE FROM A DRILLING DATA BASE

Keiji Takemura*

Institute for Geothermal Sciences, Graduate School of Science, Kyoto University, Beppu, Japan

ABSTRACT

This chapter introduces two methods for understanding the interiors of the tectonic basins and their formation history. One is a basic stratigraphic method based on the paleogeographic evidence, and the other is based on drilling information mainly in the Osaka Plain area for subsurface structures at shallow depth using a drilling database. The paleogeographic information for the sedimentary basins in the Kinki district since the Pliocene Epoch show the basin formation history related to tectonic development consistent with changing from a north-south compression to an east-west compressional stress state at the convergent margin. Information from a drilling data base, collected over a period of 40 years in this region (especially in the Osaka sedimentary basin)

* Correspondence: takemura@bep.vgs.kyoto-u.ac.jp.

indicate the distribution of marine transgressions during the Quaternary Period and a temporal subsidence mode in the central part of the Osaka sedimentary basin (Osaka Plain area).

1. Introduction

Located on the convergent margin of the Eurasian plate, the Japanese archipelago features many tectonic features and sea level changes that are well suited for studies of the Quaternary intraplate tectonics and tectono-sedimentary history (Itoh et al., 2015).

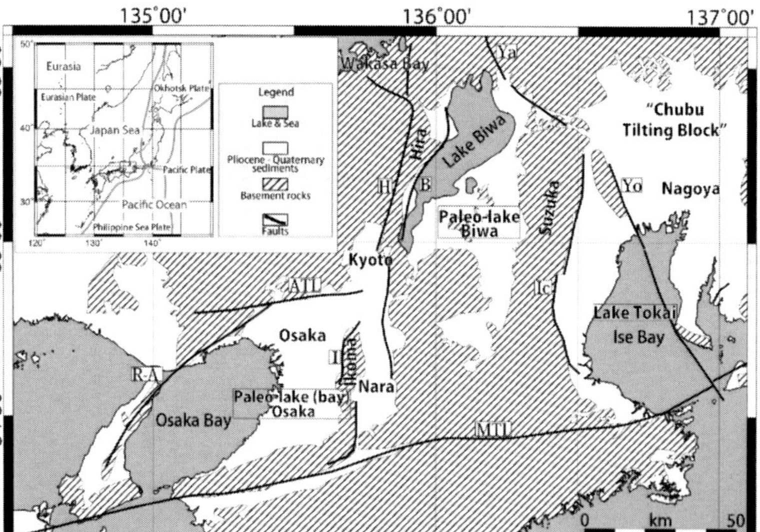

Figure 26. Tectonic setting and sedimentary basins in the Kinki district (sedimentary basin distribution with tectonic map of Japanese archipelago) (revised from Takemura et al., 2013). MTL: Median Tectonic Line, ATL: Arima-Takatsuki Tectonic Line, I: Ikoma Fault, H: Hanaore Fault, B: Biwako-seigan Fault zone, Ya: Yanagase Fault, Ic: Ichishi Fault, Yo: Yoro Fault.

In the Kinki area of central Japan, three sedimentary basins are arranged latitudinally: the Tokai sedimentary basin (Lake Tokai), the Lake Biwa sedimentary basin (Paleo-lake Biwa) and Osaka sedimentary basin (Paleo-lake (bay) Osaka) (Figure 26). The topographical features in this district are characterized by the alignment of the basins, mountain ranges covering several tens of km in width, and active fault traces along the boundary of the basin and mountain range (Takemura et al., 2013).

Paleogeographic information and paleogeographic maps play important roles when considering the sedimentary basin formation and the tectonic stress history. In this chapter, we introduce the tectonic implications according to the paleogeographic information from the stratigraphy of the Pliocene to Quaternary sediment in the Kinki district, central Japan, and give examples of the recent tectonic activity and sea level changes on the basis of the drilling (boring) data.

2. Tectonic Implication on the Basis of Paleogeography from the Pliocene to Quaternary Sediments in Kinki District

As stated previously, three sedimentary basins since Pliocene Epoch have been aligned in the Kinki district: the Tokai sedimentary basin (Lake Tokai), the Lake Biwa sedimentary basin (Paleo-lake Biwa) and the Osaka sedimentary basin (Paleo-lake (bay) Osaka) from east to west (Figure 26). All contain the Plio-Pleistocene deposits (Figure 27). They are named Tokai, Kobiwako, and Osaka Groups, respectively.

The Tokai sedimentary basin in the eastern part of the district contains lacustrine sediments from the Tokai Group (ca. 5~6 Ma) (Figure 27). The older portion of Tokai Group is distributed at the southern and eastern part of basin according to the chronostratigraphic information. The younger portion of Tokai Group is distributed to the north and northwest after migration from the SSE to the NNW (Takemura, 1985).

Lake Biwa basin in the central Kinki District contains lacustrine and fluvial sediments from the Kobiwako Group (ca 5 Ma) (Figure 27). The earliest sediments in the Kobiwako Group developed at the southern part of the sedimentary basin. Later sediments are distributed to the north and northwest after migration from the South to the North (Yokoyama, 1969; 1984). Sedimentary record from the present Lake Biwa (active basin forming phase), and drilling in 1982 and 1983 that reached to the basement of the Mesozoic-Paleozoic sedimentary formation shows about 1 km thick sediments (Horie, 1983, 1984; Yokoyama and Takemura, 1983; Takemura, 1990) in the central part of the lake (ca 1.3 Ma) (Meyers et al., 1993; Danhara et al., 2010) (Figure 28).

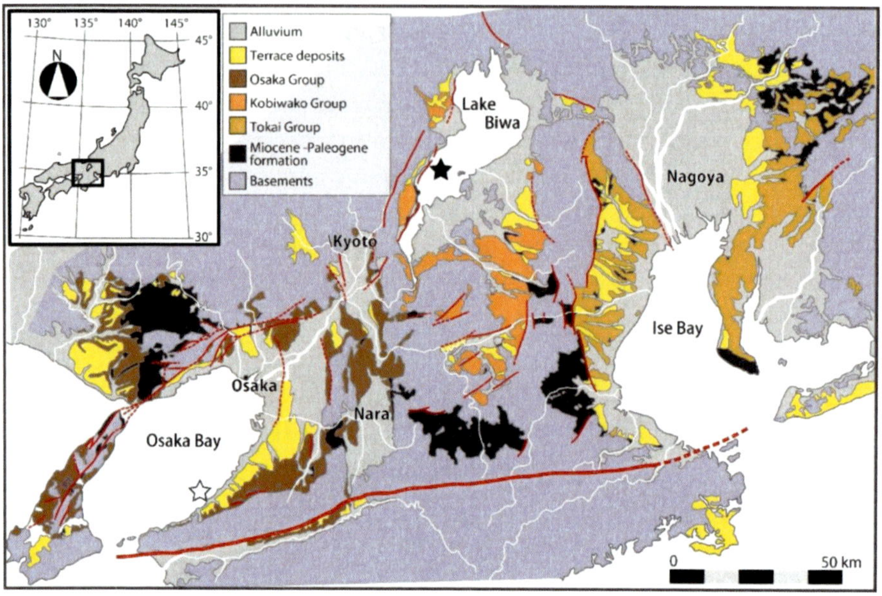

Figure 27. Distribution of sediments since the Pliocene in the Kinki district (redrawn from Takemura, 1985). Solid star: deep drilling site in 1982~1983 at Lake Biwa; open star: KIX 18-1 drilling at Kansai International Airport in 2006.

The Osaka sedimentary basin in western Kinki contains lacustrine, fluvial, and marine sediments from the Osaka Group (ca 3 Ma) (Figure 27). The upper sequence of the Osaka Group is characterized by alternating marine and non-marine sediments related to global sea level changes (Ikebe et al., 1970; Itihara, 1993). Sedimentary information is available from several deep drillings in the present Osaka sedimentary basin. The most representative is the drilling carried out in 2006 at the Kansai International Airport (KIX18-1 Editorial Committee, 2011) (Figure 29). The lower part of the sequence contains lacustrine and fluvial sediments, and the upper part of the sequence contains alternating marine and non-marine sediments. Chronological data are obtained from the tephrochronology and paleomagnetic information shown in Figure. 29. The marine clays are numbered Ma-1 to Ma13 in ascending order. The uppermost Ma13 is correlated to the Holocene transgressive sediment at marine isotope stage (MIS) 1. The clay, Ma12, indicates a marine transgression at MIS 5. The age of each marine clay bed is listed in Table 7 (Itoh et al., 2000). These marine sequences are very important tools for understanding the subsurface geology and the related tectonic subsidence or uplifting in the sedimentary basin.

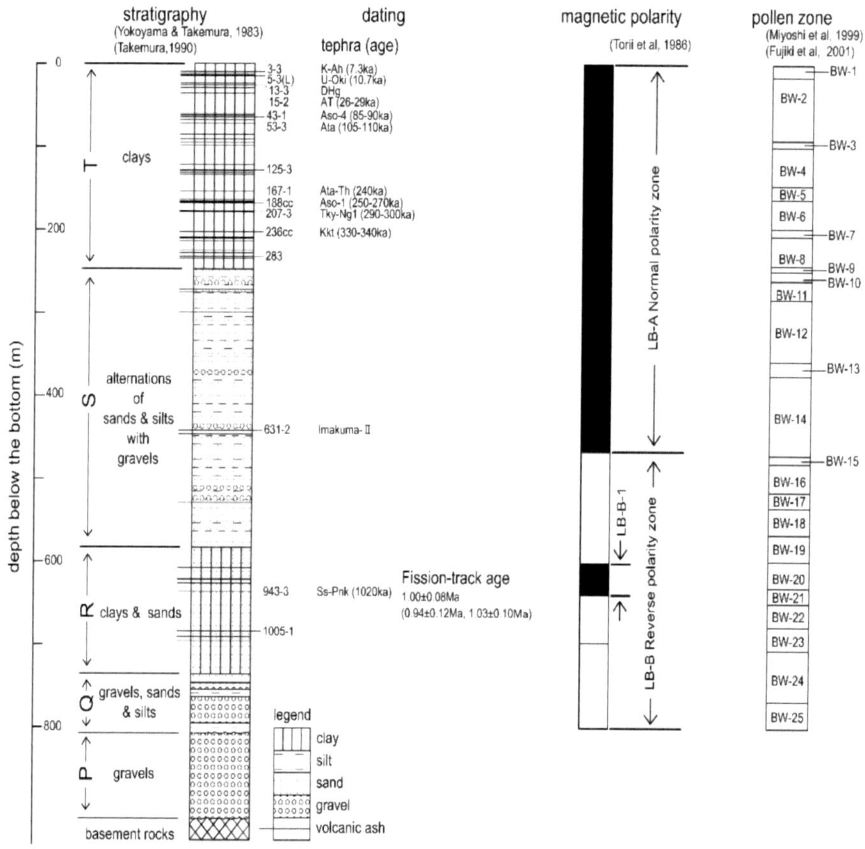

Figure 28. Stratigraphy of deep drilling carried out in 1982-1983 at present Lake Biwa basin (revised from Takemura, 1991). Left column: lithostratigraphy (Takemura, 1991), tephrostratigraphy and fission-track ages (Danhara et al., 2010), magnetostratigraphy (Torii et al., 1986) and pollen stratigraphy (Miyoshi et al, 1999 and Fujiki et al., 2001).

Paleogeographic maps from five timeframe since 5~6 Ma in the Kinki district are shown in Figure 30. Chronological correlation among the sediments from the three sedimentary basins is based on tephrochronology, dating, paleomagnetism and biostratigraphic information. The paleogeographic maps have been summarized by Yokoyama (1969), Takemura (1985), Sugiyama (1992) and Takemura et al. (2013) regarding patterns of sedimentary basin transition. On the basis of the paleogeographic map, the tectonic development in this district is suggested as follows.

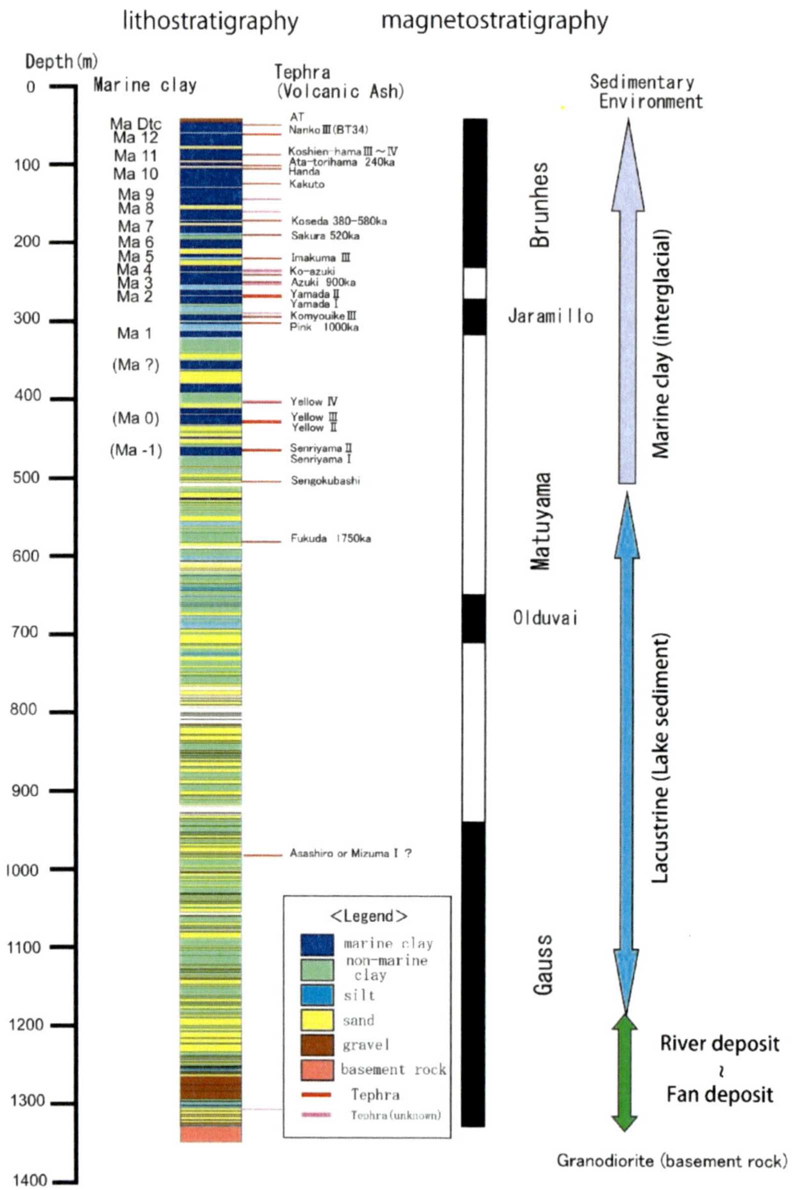

Figure 29. Deep drilling KIX 18-1 drilled in 2006 in Kansai International Airport and the stratigraphy of Osaka sedimentary basin (KIX18-1 Editorial Committee, 2011). Lithostratigraphy, tephrostratigraphy, magnetostratigraphy and environmental changes from left to right column.

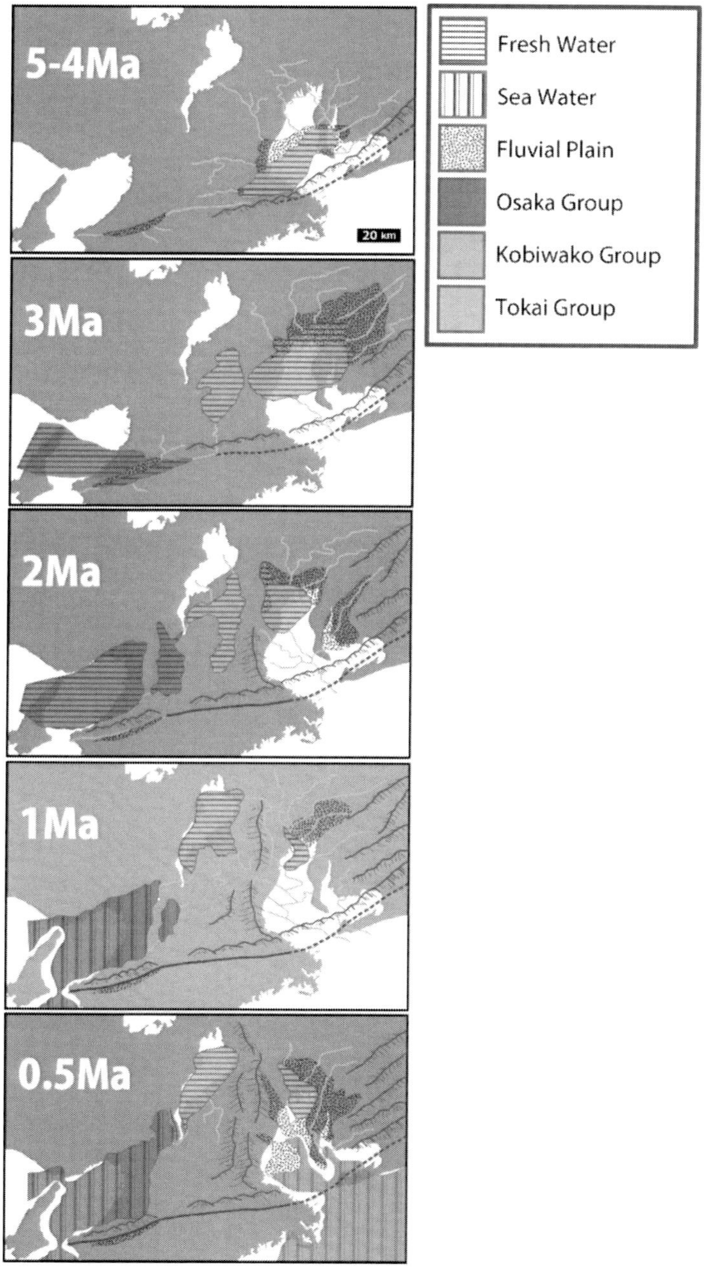

Figure 30. Paleogeographic map in the Kinki district (summarized from Yokoyama, 1969, 1984; Sugiyama, 1996; Takemura, 1985: Takemura et al., 2013). top: ca 4~5 Ma; second: ca 3Ma; third: ca 2Ma; fourth: ca 1 Ma and bottom: ca 0.5 Ma.

Table 7. Depositional age of marine clay sediments (Ma13 – Ma-1 in descending order). (Itoh et al., 2000)

Marine clay	Depositional age (10,000yr BP)
Ma13	1.1
Ma12	12.7
Ma11	24.2
Ma10	33.4
Ma9	42.7
Ma8	52.8
Ma7	57.8
Ma6	62.1
Ma5	71.2
Ma4	78.7
Ma3	86.5
Ma2	96.0
Ma1	107.0
Ma0	112.0
Ma-1	120.0

Comparing the sedimentary basin transition in Paleo-lake Biwa, Lake Tokai, and Paleo-lake (bay) Osaka, the geohistory in the central Kinki district is shown to be subdivisible by the tectono-sedimentary facies at about 3.0 Ma and 1.2~1.5 Ma, which suggests that the change in the tectonic stress state of this province took place simultaneously throughout these sedimentary basins.

According to Yokoyama (1969, 1984), the geohistory of Paleo-lake Biwa has four stages: Older I, Older II, Actual I, and Actual II. Figure 30, ca 5~4 Ma, shows the paleogeography in Older I. Figure 30, ca 3 Ma, indicates the earlier phase of Older II, Figure 30, ca 2 Ma, is the main phase of Older II. Figure 30, ca 1 Ma, shows the earlier phase of Actual I; and Figure 30, ca 0.5 Ma, indicates the early stage of Actual II that continues to present.

In Figure 30, the paleogeography at ca 5~4 Ma shows the distribution south of Lake Tokai and in the southern part of the Paleo-lake Biwa area, where clay dominant sediments were deposited in the lacustrine water body. These two basins may be connected in the earlier stage of Older I. Note that the shape of the sedimentary basin is elongated in the E-W direction.

The paleogeography at ca 3 Ma (Figure 30) shows the northward shifting of Lake Tokai and the Paleo-lake Biwa basins and the initiation of Paleo-lake Osaka. The sedimentary basin center at Paleo-lake Biwa and Lake Tokai shifted to the north, forming a stable water body with a massive clay deposition. The northward shifting of the sedimentary basin was inherited (Older II).

The paleogeography at ca 2 Ma (Figure 30) shows the distribution of sediments in the northern part of Lake Tokai and an elongated basin shape with a N-S direction in Paleo-lake Biwa. The distribution of sediments at Paleo-lake Osaka is found in the present Nara and Osaka Bay area.

The paleogeography at ca 1 Ma (Figure 30) shows the earlier stage of Actual I. During the time from Older II to Actual I, the center of the sedimentary basin migrated northwestward on a large scale. Great amounts of gravel are found as the final sediments of Older II. At the time of transition from Older II to Actual I, the Lake Tokai sedimentary basin had been diminished. The sedimentary basin of Actual I in the Lake Biwa region shifted gradually to the west, accompanied by upheaval of the eastern mountain area. In the Osaka sedimentary basin at this stage, marine transgressions have occurred since ca 1.3 Ma as recorded in the sequence from the Kansai International Airport (KIX18-1 Editorial Committee, 2011). The marine transgression reached to the eastern Nara area at that time.

The paleogeography at ca 0.5 Ma (Figure 30) shows the Actual II stage continuing through to the present active tectonic situation. The topographic contrast between the present Lake Biwa region and the western mountainous areas can be distinguished. The repeated marine transgressions related to global sea level changes reached to the Kyoto Basin and even preserved the marine sediments in the Kyoto area.

The process of basin transition in the Kinki district since the Pliocene Epoch can be summarized as follows: In the early stages, a sedimentary basin (Lake Tokai and Paleo-lake Biwa) appeared in the southern area, and migrated gradually to the north during Older I. Subsequently, the sedimentary basin migrated northwestward at a large scale during Older II. The Osaka sedimentary basin appeared at this stage. This sedimentary basin migrated gradually to the north and west. Finally, the basin was divided by structural movements with N-S trend accompanied by rapid subsidence and upheaving with N-S direction during Actual I and Actual II in the present Lake Biwa and Osaka Bay area. This resulted in conspicuous alignment of the basin and the mountains in a N-S direction.

As described previously, the geohistory of the central Kinki region had two conspicuous episodes of transition at ca 3.0 Ma and ca 1.2~1.5 Ma after the initiation of the sedimentary basin at ca 6~5 Ma. A similar pattern of basin transition was revealed in the sedimentary basins of Lake Tokai and Paleo-lake Biwa, which suggested that the changes in the tectonic stress state were common throughout that province. The sedimentary basins before ca 3.0 Ma were characterized mainly by an E-W arrangement of depression zones and their northward migration. Those characteristics might be attributable to the upheaval of the southern area under the tectonic stress state in the N-S direction. At ca 3.0 Ma, the Lake Tokai sedimentary basin migrated northwestward. The Osaka sedimentary basin was then initiated.

Subsequently, the sedimentary basin of Lake Tokai migrated to the west as a whole, although slightly northward, because of active movement in the stress state of the E-W trend. However, during the same time interval, the sedimentary basin of Paleo-lake Biwa migrated gradually northward by the upheaval of the southern area. At ca 1.2 Ma, the sedimentary basin of Lake Tokai became extinct at the time of the upheaval of the Suzuka and Yoro mountains with a N-S structural trend. It was surprising that the sedimentary basin of Paleo-lake Biwa transferred its position northwestward. Thereafter, it migrated westward by a gradual tilting of the eastern area. This indicates the origination of conspicuous movements under the E-W tectonic stress in the present Lake Biwa region. This phenomenon was observed in the Osaka sedimentary basin and accelerated from 0.5 Ma to the present.

In this way, the sedimentary basin migration in the Kinki district has been commonly explained by the interaction between the upheaval of the southern area and tilting of the eastern area. Consequently, we proposed that the deposition in the central part of the Kinki region since the Pliocene Epoch was the result of two superposed tectonic stress states represented by the upheaval of the southern area and the tilting of the eastern area. Formerly, Huzita(1969) had stated that the tectonic stress state change since the Miocene Epoch took place in the inner zone of southwestern Japan during the Pliocene and Pleistocene epochs. This change from a N-S compressional stress state to an E-W state was conspicuous. Therefore, the results of the present work did not contradict to this idea. In fact, the results confirmed the chronology and the detailed process of tectono-sedimentary turnover since the Pliocene.

3. Tectonic Implications on the Basis of Drilling Data during the Quaternary Period in the Osaka Sedimentary Basin

The data from the drilling data base are very important and are effective tools for understanding the shallow subsurface geological structure and the distribution of sediments in the Osaka Plain area (Figure 31).

Figure 31. Example of the distribution of the drilling data in Osaka City area by KG-Net drilling data base (from GI-base). Black dots indicate the drilling location.

In the Kansai (Kinki) area, a consortium of geotechnical engineers/ researchers and their affiliated organizations (the KG-NET (Kansai Geo-informatics Network)) created the geotechnical database - "Kansai Geo-Informatics Database (GI-base)". The creation of GI-base started almost 40 years ago, and the activity showed the engineering challenges, namely, the soft seabed ground, the rapid and large volume placement of near-shore reclaimed fills, the industrial subjects, etc. This collaboration between geotechnical engineering and geological science achieved during the KIA (Kansai International Airport) project and other reclamation projects advanced very the comprehensive understanding of the Osaka Bay, the Osaka Plain, and basin formation. During the activity of the KG-R (Kansai Geo-informatics - Research Committee) over the past 30 years, the Great Hanshin-Awaji Earthquake of 1995 was a monumental event that provided other collaborative research results using GI-base. The earthquake damage was spread over many structures, such as reclaimed man-made islands, harbor facilities, road/railway embankments, foundation structures, etc. In order to understand and analyze the mechanism of failures, a collaborative effort among the geotechnical engineers, the earthquake engineers, the geologists, and the structural engineers was essential. Several books were published on each sedimentary basin in the Kansai area. These were based on the collaborative researches and published at three to four year intervals using the data base of more than 60,000 drilling holes (KG-NET・Kansai Geo-informatics Research Committee, 2007) (Figure 31). In this section, the subsurface structure and their paleogeographic implications are introduced using this drilling information.

3.1. Marine Clay Distribution and Subsurface Structure: Strike and Dip of Marine Clays No. 12, 10, and 9, and the Trend of the Uemachi Flexure

As previously described, the upper sediments in the Osaka sedimentary basin are alternating marine and non-marine sediments. Each marine clay beds is named as Ma-1 to Ma13 in ascending order related to global sea level change (Table 7 and Figure 29). These marine clay sediments are important in determining the age of the sediments. The marine transgressive sediments are widely distributed over the Osaka sedimentary basin (especially in younger marine sediments). The depth of each-marine clay provides the structure of the sedimentary sequence (Inoue et al., 2013). Figure 32a and Figure 32b are examples of the subsurface structure at -25m in altitude and -45m in altitude, respectively. In those figures, the dipping direction and the dip angle of the Ma12, Ma10, and Ma9 clays are drawn. Those figures show that the dipping direction is almost ESE, and that the dip angle increases continuously in the older marine clay horizons.

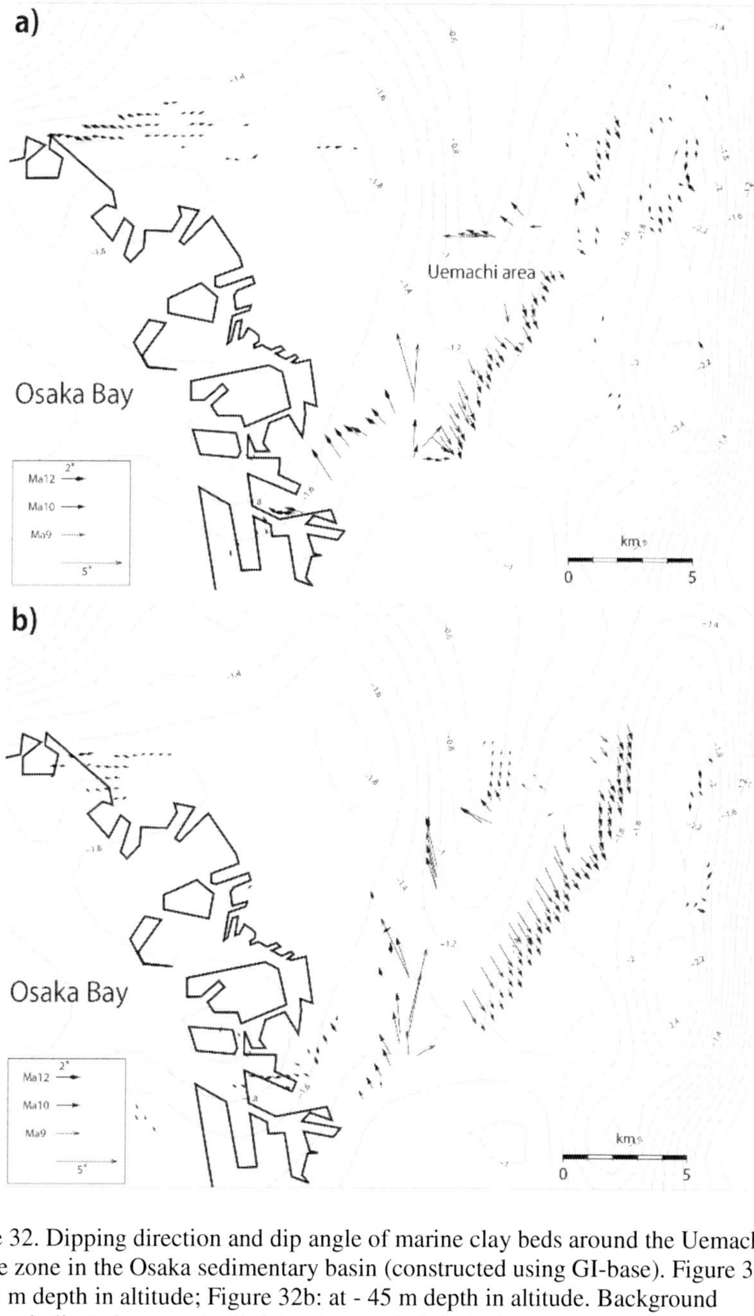

Figure 32. Dipping direction and dip angle of marine clay beds around the Uemachi flexure zone in the Osaka sedimentary basin (constructed using GI-base). Figure 32a: at - 25 m depth in altitude; Figure 32b: at - 45 m depth in altitude. Background contours indicate basement relief (contour scale is in km).

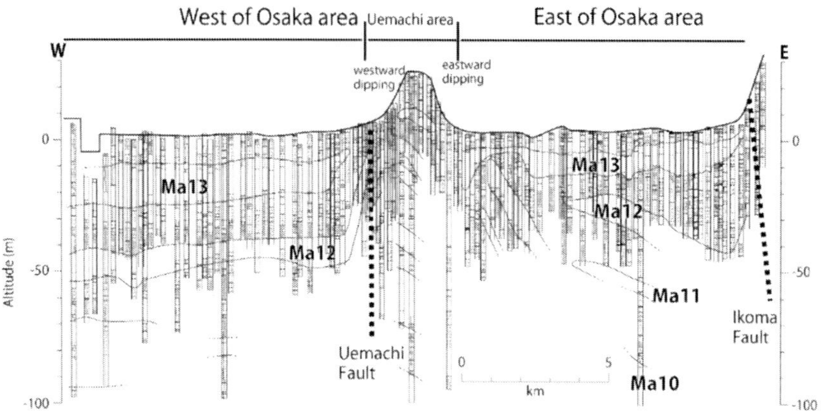

Figure 33. Example of E-W cross section of shallow stratigraphy by GI-base in Osaka Plain (redrawn from KG-Net/Kansai Geo-informatics Research Committee (2007)).

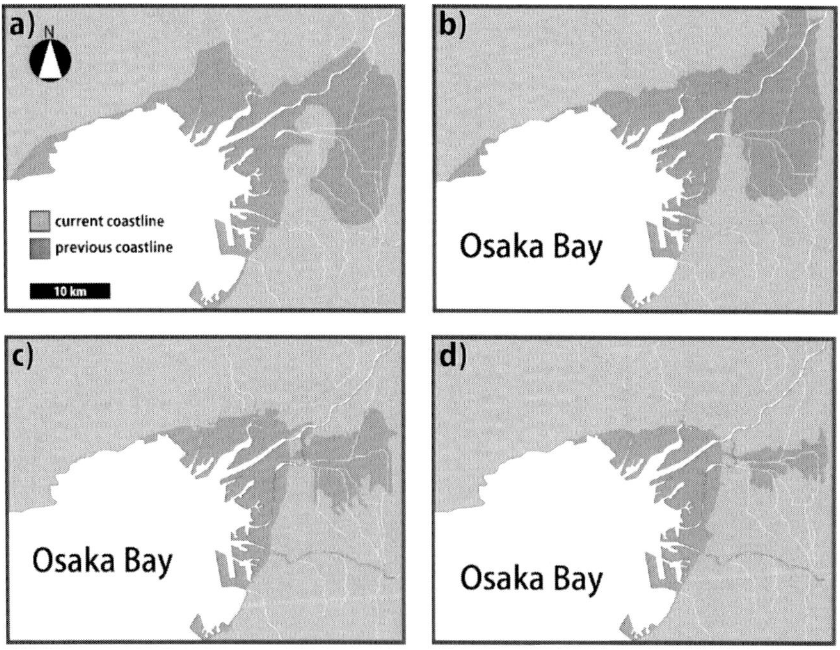

Figure 34. Paleogeographic map of the Osaka region since MIS 5 from the GI-base and archeological information. Figure 34a: map at Ma12 stage of 125 ka (MIS 5 in global cycle of sea level change), Figure 34b: map at Ma 13 stage of ca 7 ka (maximum transgression stage at Holocene (MIS 1) Figure 34c: map at ca 2 ka (stable and regression stage after rapid transgression). Figure 34d: map at ca 1.5 ka.

3.2. Paleogeography in the Osaka Sedimentary Basin since MIS 5 and the Tectonic Implications

The GI-base gave information on the precise stratigraphy and subsurface structures, especially at shallow depths. In the Figure 33 (KG-NET・Kansai Geo-informatics Research Committee, 2007), the E-W profile of the eastern part of Osaka sedimentary basin (Osaka Plain area) shows the distribution of Ma 13 and Ma 12 and two active fault traces of the Uemachi fault (west) and the Ikoma fault (east). This figure shows the asymmetrical dipping of a steep dip to the west and a gentle dip to the east at the Uemachi area. Near the Ikoma fault, Ma 12 dips gently to east and shows cumulative subsidence.

The Ma 13 (MIS 1) and Ma 12 (MIS 5) sediments are very useful for drawing the distribution of the marine transgression. Those distribution patterns have been influenced by the tectonic activity of the Ikoma fault and the Uemachi flexure area including the topographical characteristics of upheaval of the Uemachi flexure zone and cumulative subsidence along the Ikoma fault. In Figure 34a, the paleogeographic map for the Ma12 transgression during the last interglacial time (MIS 5) is shown, and wide distribution to the northern sedimentary basin and the N-S distribution between the Uemachi area and the Ikoma fault area is very noteworthy. Figure 34b is the paleogeographic map for the Holocene marine transgression and also distribution between the Uemachi area and the Ikoma fault area is very clear. Figure 34c is the paleogeographic map from the retreating stage of the coast line from about 2500 years ago, and the existence of a small bay environment can be recognized. Figure 34d is a map from the beginning of a strong human activity stage from about 1500 years ago. The area of the lagoon environment covers a narrower distribution based on the drilling data and the distribution of archaeological sites (Matsuda, 2001; Cho et al., 2013). These contrasting paleogeographic patterns must be influenced by the activity of the Uemachi flexure and the Ikoma fault.

CONCLUSION

In this chapter, as a part of this book, two methods for understanding tectonic basin formation and the basin interiors were introduced for the sedimentary basins in the Kinki District in central Japan.

The first method utilized the stratigraphy in and around the basin and the paleogeographic evidence to understand the sedimentary basin interiors and the tectonic development. The paleogeographic maps for these sedimentary basins since the Pliocene Epoch showed the basin formation history related to tectonic development and are consistent with changes from a north-south compression to an east-west compressional stress state.

The other method used drilling information mainly from the Osaka Plain area to determine subsurface structures at shallow depth. Drilling data collected during a 40 years period in this region (especially for the Osaka sedimentary basin) provided evidence that the distribution of the marine transgressions during the Quaternary Period and a temporal subsidence mode in the central part of the basin (Osaka Plain area) were related to upheaval in the central part (Uemachi area) and vertical movement of the boundary fault (Ikoma fault at the eastern end of the sedimentary basin) with a N-S direction.

ACKNOWLEDGMENTS

This paper was partially supported from Integrated research program for Uemachi Fault Zone by Ministry of Education, Sports and Culture (MEXT). Thanks for assisting figure drafting to Naoko Kitada, Naoto Inoue, Keitaro Yamada, and Miho Sasaoka.

REFERENCES

Cho, C., Ichikawa, H., Takahashi, T., Ogura, T., Hirata, Y., Matsuda, J., & Tsujimoto, Y. (2014). Topography in and around Uemachi area and the outline of paleogeographic change. In O. Wakita (Ed.), Report on JSPS Kakenhi Kiban Kenkyu A No.21242031. (in Japanese)

Danhara, T., Yamashita, T., Iwano, H., Takemura, K., & Hayashida, A. (2010). Re-investigation of chronology for the 1400m sediment core obtained from the Lake Biwa in 1982~1983. *The Quaternary Research (Japan)*, 49, 101-119. (in Japanese with English abstract)

Fujiki, T., Miyoshi, N., Morita, Y., Horie, S., & Takemura, K. (2001). Pollen analysis of an 800 m core sample from Lake Biwa, Japan. Proceedings of the IX International Palynological Congress USA (1996), Houston, Texas, 367-373.

Horie, S. (Ed.) (1983). Paleolimnology of Lake Biwa and the Japanese Pleistocene 11. Kyoto: Institute of Paleolimnology and Paleoenvironment on Lake Biwa of Kyoto University, 99pp.

Horie, S. (1991). Die Geschichte des Biwa - See in Japan: seine Entwicklung, dargestellt anhand eines 1400 m langen Tiefbohrkerns. Universitatsverlag Wagner, 349pp.

Huzita, K. (1969). Tectonic development of southwest Japan in the Quaternary Period. *Journal of Geosciences, Osaka City University*, 12, 53-70.

Ikebe, N., Iwatsu, J., & Takenaka, J. (1970). Quaternary geology of Osaka with special reference to land subsidence. *Journal of Geoscience, Osaka City University,* 13, 39-98.

Inoue, N., Kitada, N., & Takemura, K. (2013). Evaluation of Uemachi Fault zone based on borehole and underground survey data in Osaka Plain. 49th Annual Meeting of the Japanese Geotechnical Society, 299-300. (in Japanese)

Itihara, M. (Ed.) (1993). The Osaka Group. Sogensha, 340pp. (in Japanese)

Itoh, Y., Takemura, K., Ishiyama, T., Tanaka, Y., & Iwaki, H. (2000). Basin formation at a contractional bend of a large transcurrent fault: Plio-Pleistocene subsidence of the Kobe and northern Osaka Basins, Japan. *Tectonophysics,* 321, 327-341.

Itoh, Y., Kusumoto, S., & Takemura, K. (2015). Tectonically controlled asymmetric basin formation and evolution: an example from an active plate margin. In B. Veress, & J. Szigethy (Eds.), Horizons in Earth Science Research, Vol.14. New York: Nova Science Publishers, Inc., pp. 123-141. https://www.novapublishers.com/catalog/product_info.php?products_id=57536.

KG-NET/Kansai Geo-informatics Research Committee (2007). Shin Kansai Jiban; Osaka Plain to Osaka Bay. 354pp. (in Japanese)

KIX18-1 Editorial Committee (2011). Report on KIX18-1 in Kansai International Airport. 199pp. (in Japanese)

Matsuda, J. (2001). Development of tributary of Yamato River and human activity at the alluvial plain in South Kawachi Plain during the late Holocene. Abstract at 50th Research Committee of Buried Cultural Properties, 39-51. (in Japanese)

Meyers, P. A., Takemura, K., & Horie, S. (1993). Reinterpretation of Late Quaternary sediment chronology of Lake Biwa, Japan, from correlation with marine glacial - interglacial cycles. *Quaternary Research,* 39,154-162.

Miyoshi, N., Fujiki, T., & Morita, Y. (1999). Palynology of a 250-m core from Lake Biwa: A 430,000-year record of glacial-interglacial vegetation change in Japan. *Review of Palaeobotany and Palynology*, 104, 267-283, doi: 10.1016/S0034–6667(98)00058-X.

Sugiyama, Y. (1992). Neotectonics of the forarc zone and the Setouchi Province in southwest Japan. *Memoirs of the Geological Society of Japan,* 40, 219-233. (in Japanese with English abstract)

Takemura, K. (1985). The Plio-Pleistocene Tokai Group and the tectonic development around Ise Bay of central Japan since Pliocene. *Memoirs of Faculty of Science, Kyoto University, Ser. Geology & Mineralogy,* 51, 21-96.

Takemura, K. (1990). Tectonic and climatic record of the Lake Biwa, Japan, region provided by the sediments deposited since Pliocene times. *Palaeogeography, Palaeoclimatology, Palaeoecology*, 78, 185-193.

Takemura, K., Haraguchi, T., Kusumoto, S., & Itoh, Y. (2013). Tectonic basin formation in and around Lake Biwa, central Japan. In Y. Itoh (Ed.), Mechanism of Sedimentary Basin Formation - Multidisciplinary Approach on Active Plate Margins. Rijeka (Croatia): InTech. http://dx.doi.org/10.5772/56667

The Headquarter for Earthquake Research Promotion (2012). Report for Integrated Research Project on Uemachi Fault Zone. http://www.jishin.go.jp/main/chousakenkyuu/uemachi_juten/h24/index.htm delete

Torii, M., Shibuya, H., Hayashida, A., Katsura, I., Yoshida, S., Tagami, T., Otofuji, Y., Maeda, Y., Sasajima, S., & Horie, S. (1986). Magnetostratigraphy of sub-bottom sediments from Lake Biwa. *Proceedings of Japan Academy*, 62, 333-336.

Yokoyama, T. (1969). Tephrochronology and paleogeography of the Plio-Pleistocene in the eastern Setouchi Geologic province, southwest Japan. *Memoirs of Faculty of Science, Kyoto University, Geology & Mineralogy*, 36, 19-85.

Yokoyama, T. (1984). Stratigraphy of the Quaternary System around Lake Biwa and geohistory of the ancient Lake Biwa. In S. Horie (Ed.), Lake Biwa. The Hague: Dr. W. Junk, 43-128.

Yokoyama, T., & Takemura, K. (1983). Geologic column obtained by the deep drilling from the bottom surface of Lake Biwa, Japan. *IPPCCE Newsletter,* 3, 21-23.

ABOUT THE AUTHORS

Yasuto Itoh is the Professor of the Graduate School of Science of Osaka Prefecture University. He conducts research into tectonics, stratigraphy and paleomagnetism, and has published about 100 papers in the field of backarc opening processes, deformation mode of active plate margins, quantitative assessment of active faults, paleoenvironment of East Asia and mechanism of sedimentary basin formation. He is a member of the Integrated Research Project for Active Tectonics by the Ministry of Education, Culture, Sports, Science and Technology (MEXT), Japan.

Shigekazu Kusumoto was awarded MSc and PhD in geodesy from Kyoto University. He worked 8 years as an assistant professor and associate professor at the Tokai University, and he studied not only geodesy but applied geophysics and tectonophysics using numerical modeling. At present, he is an associate professor of the University of Toyama, and his present specialty is tectonophysics including applied geophysics. He has studied on formation process of sedimentary basin by dislocation modeling and discrete element modeling. And, he is interested in hydrofracture and caldera formation and studies them by theoretical approaches. In addition, he has studied on estimation techniques of subsurface structures of sedimentary basin and/or collapse caldera by gravity anomaly and gravity gradient tensor.

Keiji Takemura received a PhD in Kyoto University related to geological science. Now he is the Professor of the Graduate School of Science of Kyoto University. He conducts research into stratigraphy, active tectonics, tephrochronology and geothermal sciences, and has published more than 300 papers (half in English) in the field of Quaternary sciences of active faults, paleogeography and tectonic history of southwest Japan, environmental changes from lake and bay sediments, geothermal geology, soil engineering, and geoarchaeology. His professional memberships include: American Geophysical Union, Japan Geoscience Union, Japanese Association for Quaternary Sciences and Geological Society of Japan. He is also a leader of one program of the Integrated Research Project for Active Fault Systems by the Ministry of Education, Culture, Sports, Science and Technology (MEXT), Japan.

INDEX

A

accommodation, 4
advancement, 28
Africa, 25
age, 38, 39, 43, 46, 49, 52, 55, 58, 59, 60, 61, 62, 63, 67, 74, 78, 82
amplitude, 14, 36
Arima-Takatsuki Tectonic Line, 15, 72
assessment, 89

B

barriers, 2
base, 3, 64, 71, 80, 81, 83, 84, 85
Biwako-seigan Fault zone, 72
boreholes, 1, 4, 64, 65, 67
Butsunenjiyama Fault, 31

C

calibration, 67
case study, vii
CD-ROM, 6, 7
challenges, 81
chemical, 38
collaboration, 81
compilation, 4
compression, 66, 71, 86
compressive regime, 30, 36
compressive stress, 2, 27, 65, 67
computer, 25
configuration, 2, 3, 6, 24, 68
Congress, 86
constituents, 32
construction, 4
contour, 5, 15, 16, 64, 83
convergence, 1, 2, 6, 69
correlation, 75, 87
covering, 72
Croatia, 6, 24, 68, 69, 88
crust, 2, 5, 18, 32
crystals, 38, 59
cycles, 87

D

data collection, 25
data processing, 25
data set, 20
database, 4, 5, 6, 7, 35, 36, 71, 81
deconvolution, 9, 10, 16, 17, 18, 22, 23, 25, 26
deformation, 1, 2, 6, 31, 89
deposition, 78, 80
deposits, 30, 38, 73
depression, 4, 27, 79
depth, 14, 17, 19, 25, 30, 32, 33, 35, 71, 82, 83, 86
derivatives, 11

detection, 10, 24, 26
dimensionality, 21
Dimensionality Index, 19, 21
dip, 9, 19, 20, 21, 22, 23, 24, 82, 83, 85
Dip Angle Estimation, 19
discontinuity, 36
discordance, 41, 42, 45, 47, 48, 50, 51, 53, 54, 56, 57
dislocation, 30, 89
displacement, 32
distribution, 20, 69, 72, 78, 79, 80, 81, 85, 86
diversity, 28
dominance, 2, 66
drawing, 85
drilling (boring) data., 73
drilling data, 71, 80, 81, 85, 86
drilling database, 71, 80, 81

E

earthquakes, 2, 24
East Asia, 89
edge detection, 10, 24, 26
edge emphasis, 10, 11, 23
editors, iv
eigenvalues, 19
eigenvector, 20, 23
energy, 28, 29
engineering, 81, 90
environment, 85
environmental change, 2, 76, 90
Eötvös, 15
etching, 38
Euler deconvolution, 9, 10, 16, 17, 18, 22, 23, 25, 26
evidence, 71, 86
evolution, 28, 69, 87
exaggeration, 30, 31, 32, 33, 34
extraction, 11, 14, 16
extracts, 14, 16

F

facies, 78
filters, 24
fission, 38, 58, 59, 60, 61, 62, 63, 67, 75
fission track (FT), 38, 58, 59, 60, 61, 62, 63
flank, 66
formation, 2, 6, 24, 32, 66, 68, 69, 70, 71, 73, 81, 85, 86, 87, 88, 89
Fourier transform technique, 25
FT dating, 38, 58

G

geohistory, 78, 79, 88
geology, 65, 68, 74, 87, 90
geotechnical database, 81
gravity, 1, 3, 4, 5, 6, 9, 10, 11, 13, 14, 15, 17, 19, 21, 22, 23, 24, 25, 26, 30, 32, 33, 36, 64, 66, 68, 89
gravity gradient tensor, 9, 10, 11, 13, 15, 17, 19, 20, 21, 23, 24, 25, 89

H

history, 1, 2, 36, 71, 72, 73, 86, 90
Holocene, 74, 84, 85, 87
homogeneity, 16
horizontal differential coefficient, 10
horizontal gravity gradient, 5, 9, 11, 15, 21, 22
human, 85, 87
human activity, 85, 87

I

identification, 23
Ikoma Fault, 5, 15, 19, 22, 30, 72
Ikoma Mountains, 29, 30, 68
industrialization, 27
initiation, 78, 79
integration, 11
invariants, 19

irradiation, 67
islands, 82
isotope, 74
Izumi Mountains, 36, 37, 66

J

Japan, 1, 2, 4, 6, 7, 9, 20, 23, 24, 25, 26, 27, 29, 66, 67, 68, 69, 70, 71, 72, 73, 80, 85, 86, 87, 88, 89, 90

K

Kinki district, 71, 72, 73, 74, 75, 77, 78, 79, 80
Kobiwako Group, 73
KOH, 38
Kumedaike Fault, 33

L

Lake Biwa sedimentary basin, 72, 73
laser ablation, 28
Late Pleistocene, 64
leaching, 38

M

majority, 21
mantle, 2
mapping, 24
marine clay, 65, 66, 74, 78, 82, 83
marine isotope stage (MIS), 74, 84, 85
mass, 28
mass spectrometry, 28
materials, 11
measurement, 38
Median Tectonic Line, 20, 36, 37, 66, 67, 68, 72
migration, 1, 2, 73, 80
Ministry of Education, 86, 89, 90
Miocene, 32, 38, 80
models, 18, 28, 69
morphology, 36
mountain ranges, 72

N

nonequilibrium, 59, 61
northern sector, 30, 64, 67

O

objectivity, 10
Osaka basin, 4, 6, 30, 31, 64, 67, 68
Osaka Bay, 5, 7, 16, 19, 23, 29, 33, 35, 79, 81, 87
Osaka Group, 28, 31, 35, 38, 65, 70, 73, 74, 87
Osaka Plain, 5, 7, 16, 19, 70, 71, 80, 81, 84, 85, 86, 87
Osaka sedimentary basin, 1, 3, 4, 6, 27, 28, 30, 32, 36, 63, 64, 65, 66, 67, 71, 72, 73, 74, 76, 79, 80, 82, 83, 85, 86

P

Pacific, 2
paleogeographic maps, 73, 75, 77, 84, 85, 86
Paleogeography, 71, 73, 85
parallel, 2, 5, 20
permission, 67
Philippine Sea Plate, 2, 6, 36
physical properties, 27
Pliocene, 37, 65, 66, 70, 71, 73, 74, 79, 80, 86, 88
pollen, 75
process control, 69
project, 81
propagation, 2

R

relief, 83
researchers, 17, 81

resolution, 18, 28, 29
room temperature, 38

S

scatter, 38, 59
science, 2, 81, 89
sea level, 72, 73, 74, 79, 82, 84
sediment, 2, 28, 64, 73, 74, 86, 87
sediments, 27, 32, 37, 38, 64, 73, 74, 75, 78, 79, 80, 82, 85, 88, 90
seismic data, 28, 29, 63, 65
semi-automatic interpretation method, 9, 10, 16, 19
sensitivity, 10
shape, 36, 78, 79
shear, 2, 31
shear deformation, 2
Shobudani Group, 37, 38, 58, 59, 60, 61, 62, 63, 66
showing, 15
signals, 11, 14
solution, 18, 38
southern sector, 27, 33, 35, 36, 65, 66, 67
specifications, 28
standardization, 39, 43, 46, 49, 52, 55
state, 59, 67, 71, 78, 79, 80, 86
states, 80
stress, 2, 27, 33, 35, 36, 65, 67, 71, 73, 78, 79, 80, 86
structure, 7, 9, 11, 14, 16, 19, 20, 21, 24, 25, 32, 35, 64, 68, 80, 82
Sweden, 20, 23

T

target, 6, 38
TDX, 9, 14, 15, 16
techniques, 10, 11, 14, 23, 89
technology, 24
tectonic regime, 27, 67
tectonic stress states, 80
temperature, 58
tension, 35
Tensor Euler deconvolution, 9, 17, 18, 22
theoretical approaches, 89
thorium, 41, 42, 45, 47, 48, 50, 51, 53, 54, 56, 57
Togo, 3, 7, 33, 70
Tokai Group, 73, 88
Tokai sedimentary basin, 72, 73, 79, 80
tracks, 39, 43, 46, 49, 52, 55
transformation, 11
transgression, 74, 79, 84, 85
trial, 18
turnover, 80

U

Uemachi area, 85, 86
Uemachi Fault, 5, 15, 19, 23, 30, 64, 67, 70, 86, 87, 88
Uemachi flexure, 83, 85
Uemachi flexure area, 85
Uemachi flexure zone, 83, 85
U–Pb, 28, 38, 58, 60
U–Pb dating, 28, 58, 60
uranium, 39, 41, 42, 43, 45, 46, 47, 48, 49, 50, 51, 52, 53, 54, 55, 56, 57, 59
urban, 28
USA, 86

V

vector, 11, 13, 17, 23
vegetation, 88

W

water, 78
watershed, 2, 66
wells, 27

Y

Yamato River, 30, 33, 87
yield, 61